17th Edition
IET Wiring Regulations
Explained and Illustrated

HIT

Hertfordshire

This popular   he application
of the IET W   ge those parts
of the regulat   e correct procedures to fol   laced on areas
where confusi   s earthing and
bonding, circu   increased use
of RCDs.

With the cove   City & Guilds
Certificate in   2382-15) and
containing sa   s also an ideal
revision guide

Brian Scadda   orary Member
of City & Gui   Education and
training. He   an approved
City & Guilds   s of electrical
installation contracting including the CCG 2002-10, 2092-10, 2377-22,
2394-01, 2395-01 and 2396-01. He is also a leading author of books for
these courses.

17th Edition
IET Wiring Regulations
Explained and Illustrated

Tenth Edition

Brian Scaddan

Routledge
Taylor & Francis Group

LONDON AND NEW YORK

Tenth edition published 2015
by Routledge
2 Park Square, Milton Park, Abingdon, Oxon OX14 4RN

and by Routledge
711 Third Avenue, New York, NY 10017

Routledge is an imprint of the Taylor & Francis Group, an informa business

First edition published 1989 by Newnes, an imprint of Elsevier
Ninth edition published 2011 by Newnes, an imprint of Elsevier

British Library Cataloguing in Publication Data
A catalogue record for this book is available from the British Library

Library of Congress Cataloging in Publication Data
Scaddan, Brian.
17th edition IET wiring regulations : explained and illustrated / Brian Scaddan. – 10th edition.
pages cm
Includes index.
1. Electric wiring, Interior. 2. Electric apparatus and appliances–Installation.
3. Electric wiring–Insurance requirements–Great Britain. I. Title.
II. Title: IET wiring regulations.
TK3271.S265 2015
621.319'24--dc23
2014048213

ISBN: 978-1-138-84917-4 (pbk)
ISBN: 978-1-315-72581-9 (ebk)

Typeset in Kuenstler 480 and Trade Gothic by
Servis Filmsetting Ltd, Stockport, Cheshire

Printed by Bell and Bain Ltd, Glasgow

Contents

PREFACE vii

INTRODUCTION ix

CHAPTER 1   Fundamental Requirements for Safety 1
IET Wiring Regulations (IET Regulations Part 1 and Chapter 13) 1
The Health and Safety at Work Act 1974 3
The Electricity at Work Regulations 1989 5
The Building Regulations 2010 Part 'P' 8

CHAPTER 2   Earthing 13
Earth: What Is It, and Why and How Do We Connect to It? 15
Earth Electrode Resistance 19
Earthing in the IET Regulations (IET Regulations Chapter 4, Section 411) 22
Earthing Systems (IET Regulations Definitions (Systems)) 23
Earth Fault Loop Impedance 25
Determining the Value of Total Loop Impedance 26

CHAPTER 3   Protection 39
What is Protection? 41
Protection Against Electric Shock (IET Regulations Chapter 41) 47
SELV or PELV 50
Protection Against Overcurrent (IET Regulations Chapter 43 and Definitions) 51
Protection Against Overvoltage (IET Regulations Section 443) 59
Protection Against Undervoltage (IET Regulations Section 445) 60

CHAPTER 4   Isolation Switching and Control 61
Isolation and Switching (IET Regulations Section 537) 61

CHAPTER 5   Circuit Design 65
Design Procedure 66
Design Current 67
Nominal Setting of Protection 67
Voltage Drop (IET Regulations 525 and Appendix 12) 73
Shock Risk (IET Regulations Section 411) 74
Thermal Constraints (IET Regulations Section 543) 74
Example of Use of the Adiabatic Equation 77

An Example of Circuit Design 79
Assessment of General Characteristics 83
Sizing the Main Tails 85
Sizing the Kiln Circuit Cable...85

CHAPTER 6 Inspection and Testing..89
Testing Sequence (Part 7) 89

CHAPTER 7 Special Locations IET Regulations Part 7.....................109
Introduction 109
BS 7671 Section 701: Bathrooms, etc...110
BS 7671 Section 702: Swimming Pools 113
BS 7671 Section 703: Hot Air Saunas 115
BS 7671 Section 704: Construction Sites...116
BS 7671 Section 705: Agricultural and Horticultural Locations.............118
BS 7671 Section 706: Conductive Locations with Restrictive Movement..........120
BS 7671 Section 708: Caravan and Camping Parks.............................120
BS 7671 Section 709: Marinas...121
BS 7671 Section 710: Medical Locations...123
BS 7671 Section 711: Exhibitions, Shows and Stands..........................123
BS 7671 Section 712: Solar Photovoltaic (PV) Supply Systems 125
BS 7671 Section 714: Outdoor Lighting Installations...........................125
BS 7671 Section 715: Extra Low Voltage Lighting Installations 126
BS 7671 Section 717: Mobile or Transportable Units 126
BS 7671 Section 721: Caravans and Motor Caravans 126
BS 7671 Section 722: Electric Vehicle Charging Installations 128
BS 7671 Section 729: Operating and Maintenance Gangways 128
BS 7671 Section 740: Amusement Devices, Fairgrounds, Circuses, etc. 129
BS 7671 Section 753: Floor and Ceiling Heating Systems 129

APPENDIX A 17th Edition: Examination Question Allocation and
                Sample Paper...133
Typical City & Guilds Style Multiple Choice Question Paper 133
Sample Paper 134

APPENDIX B Answers to Multiple Choice Test 143

INDEX 145

Preface

As a result of many years developing and teaching courses devoted to compliance with the IET Wiring Regulations, it has become apparent to me that many operatives and personnel in the electrical contracting industry have forgotten the basic principles and concepts upon which electric power supply and its use are based. As a result of this, misconceived ideas and much confusion have arisen over the interpretation of the Regulations.

It is the intention of this book to dispel such misconceptions and to educate and where necessary refresh the memory of the reader. In this respect, emphasis has been placed on those areas where most confusion arises, namely earthing and bonding, protection, and circuit design.

The current seventeenth edition of the IET Wiring Regulations, also known as BS 7671, to which this book conforms, was published in January 2008 and now incorporates Amendment 3. This book is *not* a guide to the Regulations or a replacement for them; nor does it seek to interpret them Regulation by Regulation. It should, in fact, be read in conjunction with them; to help the reader, each chapter cites the relevant Regulation numbers for cross-reference.

It is hoped that the book will be found particularly useful by college students, electricians and technicians, and also by managers of smaller electrical contracting firms that do not normally employ engineers or designers. It should also be a useful addition to the library of those studying for the C&G 2382 series qualifications.

Brian Scaddan, April 2015

Introduction

It was once said, by whom I have no idea, that 'rules and regulations are for the guidance of wise men and the blind obedience of fools'. This is certainly true in the case of the IET Wiring (BS 7671) Regulations. They are not statutory rules, but recommendations for the safe selection and erection of wiring installations. Earlier editions were treated as an 'electrician's Bible': the Regulations now take the form primarily of a design document.

The IET Wiring Regulations are divided into seven parts. These follow a logical pattern from the basic requirements to inspection and testing of an installation and finally to the requirements for special locations:

Part 1 indicates the range and type of installations covered by the Regulations, what they are intended for, and the basic requirements for safety.

Part 2 is devoted to the definitions of the terms used throughout the Regulations.

Part 3 details the general information needed and the fundamental principles to be adopted before any design work can usefully proceed.

Part 4 informs the designer of the different methods available for protection against electric shock, overcurrent, etc., and how to apply those methods.

Part 5 enables the correct type of equipment, cable, accessory, etc. to be selected and erected in accordance with the requirements of Parts 1–4.

Part 6 provides details of the relevant tests to be performed on a completed installation before it is energized.

Part 7 deals with particular requirements for special installations and locations such as bathrooms, swimming pools, construction sites, etc.

Appendices. These provide tabulated and other background information required by the designer/installer/tester.

It must be remembered that the Regulations are not a collection of unrelated statements each to be interpreted in isolation; there are many cross-references throughout which may render such an interpretation valueless.

In using the Regulations I have found the index an invaluable starting place when seeking information. However, one may have to try different combinations of wording in order to locate a particular item. For example, determining how often an RCD should be tested via its test button could prove difficult since no reference is made under 'Residual current devices' or 'Testing'; however, 'Periodic testing' leads to Regulation 514.12, and the information in question is found in 514.12.2. In the index, this Regulation is referred under 'Notices'.

Fundamental Requirements for Safety

Important terms/topics covered in this chapter:

- The Health and Safety at Work Act 1974
- The Electricity at Work Regulations 1989
- The Building Regulations 2010 Part 'P'
- Distribution Network Operator DNO
- DNO supply voltages and BS 7671 voltage bands
- The Electricity Safety, Quality and Safety Regulations 2002

By the end of this chapter the reader should:

- be aware of the requirements of Part1 and Chapter 13 of BS 7671: 2008,
- understand their responsibilities under the H & S at Work Act and the EAWR,
- be able to demonstrate a knowledge and understanding of the requirements of Part 'P',
- know the standard DNO's supply voltages and tolerances,
- know the BS 7671 voltage bands.

IET WIRING REGULATIONS (IET REGULATIONS PART 1 AND CHAPTER 13)

It does not require a degree in electrical engineering to realize that electricity at *low* voltage can, if uncontrolled, present a serious threat of injury to persons or livestock, or damage to property by fire.

Clearly the type and arrangement of the equipment used, together with the quality of workmanship provided, will go a long way to minimizing danger. The following is a list of basic requirements:

1. Use good workmanship.
2. Use approved materials and equipment.

17th Edition IET Wiring Regulations: Explained and Illustrated. 978-1-138-84917-4.
© Brian Scaddan. Published by Taylor & Francis. All rights reserved.

3. Ensure that the correct type, size and current-carrying capacity of cables are chosen.

4. Ensure that equipment is suitable for the maximum power demanded of it.

5. Make sure that conductors are insulated, and sheathed or protected if necessary, or are placed in a position to prevent danger.

6. Joints and connections should be properly constructed to be mechanically and electrically sound.

7. Always provide overcurrent protection for every circuit in an installation (the protection for the whole installation is usually provided by the Distribution Network Operator [DNO]), and ensure that protective devices are suitably chosen for their location and the duty they have to perform.

8. Where there is a chance of metalwork becoming live owing to a fault, it should be earthed, and the circuit concerned should be protected by an overcurrent device or a residual current device (RCD).

9. Ensure that all necessary bonding of services is carried out.

10. Do not place a fuse, a switch or a circuit breaker, unless it is a linked switch or circuit breaker, in an earthed neutral conductor. The linked type must be arranged to break all the line conductors.

11. All single-pole switches must be wired in the line conductor only.

12. A readily accessible and effective means of isolation must be provided, so that all voltage may be cut off from an installation or any of its circuits.

13. All motors must have a readily accessible means of disconnection.

14. Ensure that any item of equipment which may normally need operating or attending by persons is accessible and easily operated.

15. Any equipment required to be installed in a situation exposed to weather or corrosion, or in explosive or volatile environments, should be of the correct type for such adverse conditions.

16. Before adding to or altering an installation, ensure that such work will not impair any part of the existing installation and that the existing is in a safe condition to accommodate the addition.

17. After completion of an installation or an alteration to an installation, the work must be inspected and tested to ensure, as far as

reasonably practicable, that the fundamental requirements for safety have been met.

These requirements form the basis of the IET Regulations.

It is interesting to note that, whilst the Wiring Regulations are not statutory, they may be used to claim compliance with Statutory Regulations such as the Electricity at Work Regulations, the Health and Safety at Work Act and Part 'P' of the Building Regulations. In fact, the Health and Safety Executive produces guidance notes for installations in such places as schools and construction sites. The contents of these documents reinforce and extend the requirements of the IET Regulations. Extracts from the Health and Safety at Work Act, the Electricity at Work Regulations and Part 'P' of the Building Regulations are reproduced below.

THE HEALTH AND SAFETY AT WORK ACT 1974

Duties of employers

Employers must safeguard, as far as is reasonably practicable, the health, safety and welfare of all the people who work for them. This applies in particular to the provision and maintenance of safe plant and systems of work, and covers all machinery, equipment and appliances used.

Some examples of the matters which many employers need to consider are:

1. Is all plant up to the necessary standards with respect to safety and risk to health?
2. When new plant is installed, is latest good practice taken into account?
3. Are systems of work safe? Thorough checks of all operations, especially those operations carried out infrequently, will ensure that danger of injury or to health is minimized. This may require special safety systems, such as 'permits to work'.
4. Is the work environment regularly monitored to ensure that, where known toxic contaminants are present, protection conforms to current hygiene standards?

5. Is monitoring also carried out to check the adequacy of control measures?
6. Is safety equipment regularly inspected? All equipment and appliances for safety and health, such as personal protective equipment, dust and fume extraction, guards, safe access arrangement, monitoring and testing devices, need regular inspection (Section 2(1) and 2(2) of the Act).

No charge may be levied on any employee for anything done or provided to meet any specific requirement for health and safety at work (Section 9).

Risks to health from the use, storage, or transport of 'articles' and 'substances' must be minimized. The term *substance* is defined as 'any natural or artificial substance whether in solid or liquid form or in the form of gas or vapour' (Section 53(1)).

To meet these aims, all reasonably practicable precautions must be taken in the handling of any substance likely to cause a risk to health. Expert advice can be sought on the correct labelling of substances, and the suitability of containers and handling devices. All storage and transport arrangements should be kept under review.

Safety information and training

It is now the duty of employers to provide any necessary information and training in safe practices, including information on legal requirements.

Duties to others

Employers must also have regard for the health and safety of the self-employed or contractors' employees who may be working close to their own employees, and for the health and safety of the public who may be affected by their firm's activities.

Similar responsibilities apply to self-employed persons, manufacturers and suppliers.

Duties of employees

Employees have a duty under the Act to take reasonable care to avoid injury to themselves or to others by their work activities, and to cooperate with employers and others in meeting statutory requirements. The Act also requires employees not to interfere with or misuse anything provided to protect their health, safety or welfare in compliance with the Act.

THE ELECTRICITY AT WORK REGULATIONS 1989

Persons on whom duties are imposed by these Regulations

1. Except where otherwise expressly provided in these Regulations, it shall be the duty of every:
   a. employer and self-employed person to comply with the provisions of these Regulations in so far as they relate to matters which are within his or her control; and
   b. manager of a mine or quarry (within in either case the meaning of Section 180 of the Mines and Quarries Act 1954) to ensure that all requirements or prohibitions imposed by or under these Regulations are complied with in so far as they relate to the mine or quarry or part of a quarry of which he or she is the manager and to matters which are within his or her control.
2. It shall be the duty of every employee while at work:
   a. to cooperate with his or her employer in so far as is necessary to enable any duty placed on that employer by the provisions of these Regulations to be complied with; and
   b. to comply with the provisions of these Regulations in so far as they relate to matters which are within his or her control.

Employer

1. For the purposes of the Regulations, an employer is any person or body who (a) employs one or more individuals under a contract of employment or apprenticeship; or (b) provides training under the schemes to which the HSW Act applies through the Health and Safety (Training for Employment) Regulations 1988 (Statutory Instrument No. 1988/1222).

Self-employed

2. A self-employed person is an individual who works for gain or reward otherwise than under a contract of employment whether or not he or she employs others.

Employee

3. Regulation 3(2)(a) reiterates the duty placed on employees by Section 7(b) of the HSW Act.
4. Regulation 3(2)(b) places duties on employees equivalent to those placed on employers and self-employed persons where these are matters within their control. This will include those trainees who will be considered as employees under the Regulations described in paragraph 1.
5. This arrangement recognizes the level of responsibility which many employees in the electrical trades and professions are expected to take on as part of their job. The 'control' which they exercise over the electrical safety in any particular circumstances will determine to what extent they hold responsibilities under the Regulations to ensure that the Regulations are complied with.
6. A person may find him or herself responsible for causing danger to arise elsewhere in an electrical system, at a point beyond their own installation. This situation may arise, for example, due to unauthorized or unscheduled back feeding from the installation onto the system, or to raising the fault power level on the system above rated and agreed maximum levels due to connecting extra generation capacity, etc. Because such circumstances are 'within his or her control', the effect of Regulation 3 is to bring responsibilities for compliance with the rest of the regulations to that person, thus making him or her a duty holder.

Absolute/reasonably practicable

7. Duties in some of the Regulations are subject to the qualifying term 'reasonably practicable'. Where qualifying terms are absent the requirement in the Regulation is said to be absolute. The meaning

of reasonably practicable has been well established in law. The interpretations below are given only as a guide to duty holders.

Absolute

8. If the requirement in a Regulation is 'absolute', for example if the requirement is not qualified by the words 'so far as is reasonably practicable', the requirement must be met regardless of cost or any other consideration. Certain of the regulations making such absolute requirements are subject to the Defence provision of Regulation 29.

Reasonably practicable

9. Someone who is required to do something 'so far as is reasonably practicable' must assess, on the one hand, the magnitude of the risks of a particular work activity or environment and, on the other hand, the costs in terms of the physical difficulty, time, trouble and expense which would be involved in taking steps to eliminate or minimize those risks. If, for example, the risks to health and safety of a particular work process are very low, and the cost or technical difficulties of taking certain steps to prevent those risks are very high, it might not be reasonably practicable to take those steps. The greater the degree of risk, the less weight that can be given to the cost of measures needed to prevent that risk.

10. In the context of the Regulations, where the risk is very often that of death, for example from electrocution, and where the nature of the precautions which can be taken are so often very simple and cheap, e.g. insulation, the level of duty to prevent that danger approaches that of an absolute duty.

11. The comparison does not include the financial standing of the duty holder. Furthermore, where someone is prosecuted for failing to comply with a duty 'so far as is reasonably practicable', it would be for the accused to show the court that it was not reasonably practicable for him or her to do more than he or she had in fact done to comply with the duty (Section 40 of the HSW Act).

THE BUILDING REGULATIONS 2010 PART 'P'

Part 'P' of the building regulations requires that installations in dwellings be designed, installed and inspected and tested to the requirements of the current edition of the IET Wiring Regulations, BS 7671.

Details of some such work need to be notified to the Local Authority Building Control (LABC). This is a legal requirement.

Notifiable work

- Any new circuit fed from the consumer unit.
- Replacement of a consumer unit.
- Any work, additions or alterations in the bath/shower room, sauna or swimming pool.

Non-notifiable work

Any other work that involves replacements, additions or alterations anywhere else including gardens.

Certification

All electrical work has to be inspected, tested and certified with the issue of either:

- An Electrical Installation Certificate (EIC) for all new circuits and consumer unit replacements.
- A Minor Electrical Installation Works Certificate (MEIWC) for small additions and/or alterations that do not comprise a new circuit.
- An Electrical Installation Condition Report (EICR) where the inspector is *not* the installer.

If an installer belongs to a 'Competent Persons' scheme, i.e. is a Domestic Installer with an approval body such as the NICEIC, ELECSA etc., he/she can self-certify the work and issue the appropriate certificate to the person ordering the work. The approval body will notify the LABC.

If an installer is not registered on a 'Competent Persons' scheme, but is qualified to inspect and test, he/she should issue the appropriate certificate to the LABC. They will take the inspector's qualifications into account and decide if any further action needs to be taken.

An installer who is not qualified to inspect and test may appoint a 'Registered Competent Person' to carry out the inspection and test. In this case only an EICR can be issued to the person ordering the work.

Appendix 2 of the IET Regulations lists all of the other Statutory Regulations and Memoranda with which electrical installations must comply.

It is interesting to note that if an installation fails to comply with Chapter 13 of the Regulations, the DNO has the right to refuse to give a supply or, in certain circumstances, to disconnect it.

While we are on the subject of DNOs, let us look at the current declared supply voltages and tolerances. In order to align with European Harmonized Standards, our historic 415 V/240 V declared supply voltages have now become 400 V/230 V. However, this is only a paper exercise, and it is unlikely that consumers will notice any difference for many years, if at all. Let me explain, using single-phase as the example.

The supply industry declared voltage was 240 V ± 6%, giving a range between 225.6 V and 254.4 V. The new values are 230 V + 10% − 6%, giving a range between 216.2 V and 253 V. Not a lot of difference. The industry has done nothing physical to reduce voltages from 240 V to 230 V, it is just the declaration that has been altered. Hence a measurement of voltage at supply terminals will give similar readings to those we have always known. Figure 1.1 shows the UK supply system and associated declared voltages.

BS 7671 details two voltage categories, Band 1 and Band 2. Band 1 is essentially extra low-voltage (ELV) systems and Band 2 low-voltage (LV) systems.

ELV is less than 50 V a.c. between conductors or to earth. LV exceeds ELV up to 1000 V a.c. between conductors and 600 V between conductors and earth.

The suppliers are now governed by the 'Electricity Safety, Quality & Continuity Regulations 2002' ESQCR (formerly the Electricity Supply Regulations 1988).

Note: The connection of the transformer star or neutral point to earth helps to maintain that point at or very near zero volts.

FIGURE 1.1 DNO supply voltages.

Questions

1. Which Part of BS 7671 forms its basis?
2. Which three sets of legislation are most likely to be associated with BS 7671?
3. Under the EAWR 1989, which persons have a duty of care placed upon them?
4. Which item of legislation is solely applicable to electrical installations in dwellings?
5. When may an unqualified installer issue an Electrical Installation Certificate?
6. Under Part 'P' of the Building Regulations, to whom should work in a bathroom be notified?
7. Where, in BS 7671:2008, are Statutory Documents listed?
8. What are the DNO's standard voltages and tolerances?
9. What is the voltage Band and maximum a.c. value of Extra Low Voltage?
10. What is the ESQCR?

Answers

1. Part 1
2. The H&S at Work Act 1974, the EAWR 1989, and Part'P' of the Building Regulations
3. Employers, self employed, employees and managers of mines and quarries
4. Approved Document P of the Building Regulations
5. Never
6. The LABC
7. Appendix 2
8. 230V, 400V; +10%, −6%
9. Band 1; 50V a.c.
10. The Electricity Safety, Quality and Continuity Regulations 2002. These are applicable to the DNOs

Earthing

☞ Relevant BS 7671 chapters and parts: Chapters 31, 41, 54, Part 7

Important terms/topics covered in this chapter:

- **Basic protection** Protection against electric shock under fault-free conditions
- **Bonding conductor** A protective conductor providing equipotential bonding
- **Circuit protective conductor (cpc)** A protective conductor connecting exposed conductive parts of equipment to the main earthing terminal
- **Earth** The conductive mass of earth, whose electric potential at any point is conventionally taken as zero
- **Earth electrode resistance** The resistance of an earth electrode to earth
- **Earth fault current** An overcurrent resulting from a fault of negligible impedance between a line conductor and an exposed conductive part or a protective conductor
- **Earth fault loop impedance** The impedance of the phase-to-earth loop path starting and ending at the point of fault
- **Earthing conductor** A protective conductor connecting a main earthing terminal of an installation to an earth electrode or other means of earthing
- **Equipotential bonding** Electrical connection maintaining various exposed conductive parts and extraneous conductive parts at a substantially equal potential
- **Exposed conductive part** A conductive part of equipment which can be touched and which is not a live part but which may become live under fault conditions
- **Extraneous conductive part** A conductive part liable to introduce a potential, generally earth potential, and not forming part of the electrical installation

17th Edition IET Wiring Regulations: Explained and Illustrated. 978-1-138-84917-4.

- **Fault protection** Protection against electric shock under single-fault conditions
- **Functional earth** Earthing of a point or points in a system or an installation or in equipment for purposes other than safety, such as for proper functioning of electrical equipment
- **Leakage current** Electric current in an unwanted conductive part under normal operating conditions
- **Line conductor** A conductor of an a.c. system for the transmission of electrical energy, other than a neutral conductor
- **Live part** A conductor or conductive part intended to be energized in normal use, including a neutral conductor but, by convention, not a PEN conductor
- **PEN conductor** A conductor combining the functions of both protective conductor and neutral conductor
- **PME (protective multiple earthing)** An earthing arrangement, found in TN-C-S systems, where an installation is earthed via the supply neutral conductor
- **Protective conductor** A conductor used for some measure of protection against electric shock and intended for connecting together any of the following parts:
  - exposed conductive parts
  - extraneous conductive parts
  - main earthing terminal
  - earth electrode(s)
  - earthed point of the source
- **Residual current device (RCD)** An electromechanical switching device or association of devices intended to cause the opening of the contacts when the residual current attains a given value under given conditions
- **Simultaneously accessible parts** Conductors or conductive parts which can be touched simultaneously by a person or, where applicable, by livestock

By the end of this chapter the reader should:

- know why the UK distribution is earthed,
- know why a consumer's connection to earth is important,
- be aware of the methods of connecting to earth,

- understand the importance of earth electrode resistance,
- know why it is advantageous to enclose an earth rod below ground level,
- understand the difference between and the importance of earthing and bonding,
- know the different between and the arrangements of supply earthing arrangements,
- know the component parts of the earth fault loop path,
- be able to determine values of earth fault loop impedance,
- know the ratings of RCDs for various applications,
- understand the operation of RCDs,
- know when supplementary equipotential bonding is required.

EARTH: WHAT IS IT, AND WHY AND HOW DO WE CONNECT TO IT?

The thin layer of material which covers our planet, be it rock, clay, chalk or whatever, is what we in the world of electricity refer to as earth. So, why do we need to connect anything to it? After all, it is not as if earth is a good conductor.

Perhaps it would be wise at this stage to investigate potential difference (PD). A PD is exactly what it says it is: a difference in potential (volts). Hence, two conductors having PDs of, say, 20 V and 26 V have a PD between them of $26 - 20 = 6$ V. The original PDs, i.e. 20 V and 26 V, are the PDs between 20 V and 0 V and 26 V and 0 V.

So where does this 0 V or zero potential come from? The simple answer is, in our case, the earth. The definition of earth is therefore the conductive mass of earth, whose electric potential at any point is conventionally taken as zero.

In the UK, the distribution system has the Neutral of the supply transformer connected to earth in order to maintain it at as near to zero volts as possible and to reduce the risk of shock at 400 V in the event of earth faults.

Hence, if we connect a voltmeter between a live part (e.g. the line conductor of, say, a socket outlet circuit) and earth, we may read 230 V; the conductor is at 230 V, the earth at zero. In the UK, the distribution

FIGURE 2.1   (a) Earth path and (b) no earth path.

system has the Neutral of the supply transformer connected to earth in order to maintain it at as near zero volts as possible and to reduce the risk of shock at 400 V in the event of earth faults. The earth provides a path to complete the circuit, via the earthed neutral of the supply transformer. We would measure nothing at all if we connected our voltmeter between, say, the positive 12 V terminal of a car battery and earth, as in this case the earth plays no part in any circuit. Figure 2.1 illustrates this difference.

FIGURE 2.2   Shock path.

Hence, a person in an installation touching a live part whilst standing on the earth would take the place of the voltmeter in Figure 2.1a, and could suffer a severe electric shock. Remember that the accepted lethal level of shock current passing through a person is only 50 mA or 1/20 A. The same situation would arise if the person were touching, say, a faulty appliance and a gas or water pipe (Figure 2.2).

One method of providing some measure of protection against these effects is to join together (bond) all metallic parts and connect them to earth. This ensures that all metalwork in a healthy situation is at or near zero volts, and under fault conditions all metalwork will rise to a similar potential. So, simultaneous contact with two such metal parts would not result in a dangerous shock, as there will be no significant PD between them. This method is known as protective equipotential bonding.

Unfortunately, as previously mentioned, earth itself is not a good conductor unless it is very wet, and therefore it presents a high resistance to the flow of fault current. This resistance is usually enough to restrict fault current to a level well below that of the rating of the protective device, leaving a faulty circuit uninterrupted. Clearly this is an unhealthy situation. The methods of overcoming this problem will be dealt with later.

In all but the most rural areas, consumers can connect to a metallic earth return conductor which is ultimately connected to the earthed neutral of the supply. This, of course, presents a low-resistance path for fault currents to operate the protection.

Summarizing, then, connecting metalwork to earth places that metal at or near zero potential, and bonding between metallic parts puts such parts at a similar potential even under fault conditions.

Connecting to earth

In the light of previous comments, it is obviously necessary to have as low an earth path resistance as possible, and the point of connection to earth is one place where such resistance may be reduced. When two conducting surfaces are placed in contact with each other, there will be a resistance to the flow of current dependent on the surface areas in contact. It is clear, then, that the greater surface contact area with earth that can be achieved, the better.

There are several methods of making a connection to earth, including the use of rods, plates and tapes. By far the most popular method in everyday use is the rod earth electrode. The plate type needs to be buried at a sufficient depth to be effective and, as such plates may be 1 or $2\,m^2$, considerable excavation may be necessary. The tape type is predominantly used in the earthing of large electricity substations, where the tape is laid in trenches in a mesh formation over the whole site. Items of plant are then earthed to this mesh.

Rod electrodes

These are usually of solid copper or copper-clad carbon steel, the latter being used for the larger-diameter rods with extension facilities. These facilities comprise: a thread at each end of the rod to enable a coupler to be used for connection of the next rod; a steel cap to protect the thread from damage when the rod is being driven in; a steel driving tip; and a clamp for the connection of an earth tape or conductor (Figure 2.3).

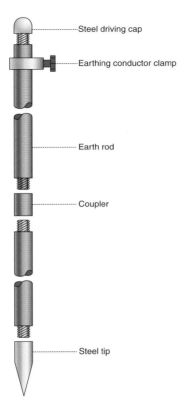

FIGURE 2.3 Earth rod.

The choice of length and diameter of such a rod will, as previously mentioned, depend on the soil conditions. For example, a long thick electrode is used for earth with little moisture retention. Generally, a 1–2 m rod, 16 mm in diameter, will give a relatively low resistance.

EARTH ELECTRODE RESISTANCE

If we were to place an electrode in the earth and then measure the resistance between the electrode and points at increasingly larger distances

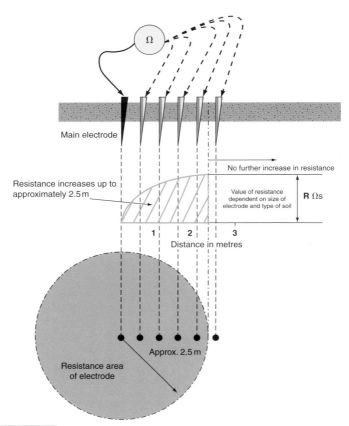

Main electrode

Resistance increases up to
approximately 2.5 m

No further increase in resistance

Value of resistance
dependent on size of
electrode and type of soil

R Ωs

1   **2**   3
Distance in metres

Approx. 2.5 m

Resistance area
of electrode

FIGURE 2.4   Earth electrode resistance area.

from it, we would notice that the resistance increased with distance until
a point was reached (usually around 2.5 m) beyond which no increase in
resistance was noticed (Figure 2.4).

The resistance area around the electrode is particularly important with
regard to the voltage at the surface of the ground (Figure 2.5, see page
21). For a 2 m rod, with its top at ground level, 80–90% of the voltage
appearing at the electrode under fault conditions is dropped across the

FIGURE 2.5 Electrode resistance.

FIGURE 2.6 Earth electrode installation.

earth in the first 2.5–3 m. This is particularly dangerous where livestock is present, as the hind and fore legs of an animal can be respectively inside and outside the resistance area: a PD of 25 V can be lethal! One method of overcoming this problem is to house the electrode in a pit below ground level (Figure 2.6) as this prevents voltages appearing at ground level. This arrangement will also provide protection for the earth connection against mechanical damage and corrosion.

EARTHING IN THE IET REGULATIONS (IET REGULATIONS CHAPTER 4, SECTION 411)

In the preceding pages we have briefly discussed the reasons for, and the importance and methods of, earthing. Let us now examine the subject in relation to the IET Regulations.

Contact with metalwork made live by a fault is clearly undesirable. One popular method of providing some measure of protection against the effects of such contact is by protective earthing, protective equipotential bonding and automatic disconnection in the event of a fault. This entails the connection to earth of:

1. All metalwork associated with electrical systems that are not live parts but which could become live under fault conditions. These are termed exposed conductive parts. This is Earthing.
2. The joining together and connection to earth of all metalwork liable to introduce a potential including earth potential, termed extraneous conductive parts. Examples are gas, oil and water pipes, structural steelwork, radiators and baths. This is Bonding.

The conductors used in such connections are called *protective conductors*, and they can be further subdivided into:

1. Circuit protective conductors, for connecting exposed conductive parts to the main earthing terminal.
2. Main protective bonding conductors, for bonding together main incoming services, structural steelwork, etc.
3. Supplementary bonding conductors for bonding exposed conductive parts and extraneous conductive parts, when circuit disconnection times cannot be met, or in special locations, such as bathrooms, swimming pools, etc.

The effect of all this earthing and bonding is to create a zone in which all metalwork of different services and systems will, even under fault conditions, be at a substantially equal potential. If, added to this, there is a low-resistance earth return path, the protection should operate fast enough to prevent danger (IET Regulations 411.3–411.6).

The resistance of such an earth return path will depend upon the system (see the next section), either TT, TN-S or TN-C-S (IT systems will not be discussed here, as they are extremely rare and unlikely to be encountered by the average contractor).

EARTHING SYSTEMS (IET REGULATIONS DEFINITIONS (SYSTEMS))

These have been designated in the IET Regulations using the letters T, N, C and S. These letters stand for:

T	terre (French for earth) and meaning a direct connection to earth
N	neutral
C	combined
S	separate

When these letters are grouped they form the classification of a type of system. The first letter in such a classification denotes how the supply source is earthed. The second denotes how the metalwork of an installation is earthed. The third and fourth indicate the functions of neutral and protective conductors. Hence:

1. A TT system has a direct connection of the supply source neutral to earth and a direct connection of the installation metalwork to earth. An example is an overhead line supply with earth electrodes, and the mass of earth as a return path (Figure 2.7).
2. A TN-S system has the supply source neutral directly connected to earth, the installation metalwork connected to the earthed neutral of the supply source via the lead sheath of the supply cable, and the neutral and protective conductors throughout the whole system performing separate functions (Figure 2.8).
3. A TN-C-S system is as the TN-S but the supply cable sheath is also the neutral, i.e. it forms a combined earth/neutral conductor known as a PEN (protective earthed neutral) conductor (Figure 2.9).

The installation earth and neutral are separate conductors. The DNO's part of this system is known as PME.

Note that only single-phase systems have been shown, for simplicity.

FIGURE 2.7 TT system.

FIGURE 2.8 TN-S system.

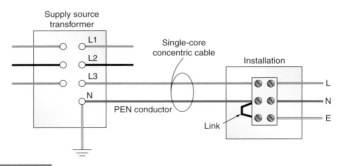

FIGURE 2.9 TN-C-S system.

Summary

In order to reduce the risk of serious electric shock, it is important to provide a path for earth fault currents to operate the circuit protection, and to endeavour to maintain all metalwork at a substantially equal potential. This is achieved by bonding together metalwork of electrical and non-electrical systems to earth. The path for earth fault currents would then be via the earth itself in TT systems or by a metallic return path in TN-S or TN-C-S systems.

EARTH FAULT LOOP IMPEDANCE

As we have seen, circuit protection should operate in the event of a direct fault from line to earth (automatic disconnection). The speed of operation of the protection is extremely important and will depend on the magnitude of the fault current, which in turn will depend on the impedance of the earth fault loop path, Z_s.

Figure 2.10 shows this path. Starting at the fault, the path comprises:

1. The cpc.
2. The consumer's earthing terminal and earthing conductor.
3. The return path, either metallic or earth, dependent on the earthing system.
4. The earthed neutral of the supply transformer.
5. The transformer winding.
6. The line conductor from the transformer to the fault.

Figure 2.11 is a simplified version of this path. We have:

$$Z_s = Z_e + R_1 + R_2$$

where Z_s is the actual total loop impedance, Z_e is the loop impedance external to the installation, R_1 is the resistance of the line conductor, and R_2 is the resistance of the cpc. We also have:

$$I = \frac{U_0}{Z_s}$$

where I is the fault current and U_0 is the nominal line voltage to earth.

FIGURE 2.10   Earth fault loop path.

FIGURE 2.11   Simplified loop path.

DETERMINING THE VALUE OF TOTAL LOOP IMPEDANCE

The IET Regulations require that when the general characteristics of an installation are assessed, the loop impedance Z_e external to the installation shall be ascertained.

FIGURE 2.12 Diagram for Example 2.2

This may be measured in existing installations using a line-to-earth loop impedance tester. However, when a building is only at the drawing board stage it is clearly impossible to make such a measurement. In this case, we have three methods available to assess the value of Z_s:

1. Determine it from details (if available) of the supply transformer, the main distribution cable and the proposed service cable; or
2. Measure it from the supply intake position of an adjacent building having service cable of similar size and length to that proposed; or
3. Use maximum likely values issued by the supply authority as follows:
   TT system: $21\,\Omega$ maximum
   TN-S system: $0.80\,\Omega$ maximum
   TN-C-S system: $0.35\,\Omega$ maximum.

Method 1 will be difficult for anyone except engineers. Method 3 can, in some cases, result in pessimistically large cable sizes. Method 2, if it is possible to be used, will give a closer and more realistic estimation of Z_e. However, if in any doubt, use method 3.

Having established a value for Z_e, it is now necessary to determine the impedance of that part of the loop path internal to the installation. This is, as we have seen, the resistance of the line conductor plus the resistance of the cpc, i.e. $R_1 + R_2$. Resistances of copper conductors may be found from manufacturers' information which gives values of resistance/metre for copper and aluminium conductors at 20°C in mΩ/m. Table 2.1 gives resistance values for copper conductors up to $35\,\text{mm}^2$.

Table 2.1   Resistance of Copper Conductors at 20°C

Conductor CSA (mm²)	Resistance (mΩ/m)
1.0	18.1
1.5	12.1
2.5	7.41
4.0	4.61
6.0	3.08
10.0	1.83
16.0	1.15
25.0	0.727
35.0	0.524

A 25 mm² line conductor with a 4 mm² cpc has $R_1 = 0.727$ and $R_2 = 4.61$, giving $R_1 + R_2 = 0.727 + 4.61 = 5.337$ mΩ/m. So, having established a value for $R_1 + R_2$, we must now multiply it by the length of the run and divide by 1000 (the values given are in mΩ/m). However, this final value is based on a temperature of 20°C, but when the conductor is fully loaded its temperature will increase. In order to determine the value of resistance at conductor operating temperature, a multiplier is used.

This multiplier, applied to the 20°C value of resistance, is determined from the following formula:

$$R_t = R_{20} \left\{ 1 + \alpha_{20}(\theta - 20) \right\}$$

where

R_t = the resistance at conductor operating temperature

R_{20} = the resistance at 20°C

α_{20} = the 20°C temperature coefficient of copper, 0.004 Ω/Ω/°C

θ = the conductor operating temperature.

Clearly, the multiplier is $\left\{ 1 + \alpha_{20}(\theta - 20) \right\}$.

So, for a 70°C thermoplastic insulated conductor (Table 54C IET Regulations), the multiplier becomes:

$$\left\{ 1 + 0.004(70° - 20°) \right\} = 1.2$$

And for a 90°C XLPE type cable it becomes:

$$\{1 + 0.004(90° - 20°)\} = 1.28$$

Hence, for a 20 m length of 70°C PVC insulated 16 mm^2 line conductor with a 4 mm^2 cpc, the value of $R_1 + R_2$ would be:

$$R_1 + R_2 = \frac{[(1.15 + 4.61) \times 20 \times 1.2]}{1000} = 0.138\,\Omega$$

We are now in a position to determine the total earth fault loop impedance Z_s from:

$$Z_s = Z_e + R_1 + R_2$$

As previously mentioned, this value of Z_s should be as low as possible to allow enough fault current to flow to operate the protection as quickly as possible. Tables 41.2–41.4 of the IET Regulations give maximum values of loop impedance for different sizes and types of protection for both final circuits, and distribution circuits for TN systems.

Provided that the actual values calculated do not exceed those tabulated, final circuits up to 32 A will disconnect under earth fault conditions in 0.4 s or less, and distribution circuits and final circuits over 32 A, in 5 s or less. The reasoning behind these different times is based on the time that a faulty circuit can reasonably be left uninterrupted. Hence, socket outlet circuits from which hand-held appliances may be used clearly present a greater shock risk than distribution circuits. It should be noted that these times, i.e. 0.4 s and 5 s, do not indicate the duration that a person can be in contact with a fault. They are based on the probable chances of someone being in contact with exposed or extraneous conductive parts at the precise moment that a fault develops. For TT systems the corresponding times are 0.2 s and 5 s.

☞ See also Table 41.1 of the IET Regulations.

Example 2.1

Let us now have a look at a typical example of, say, a shower circuit run in an 18 m length of 6.0 mm^2 (6242 Y) twin cable with cpc, and protected by a 30 A BS 3036 semi-enclosed rewirable fuse. A 6.0 mm^2 twin cable has a 2.5 mm² cpc. We will also assume that the external loop impedance Z_e is measured as 0.27 Ω. Will there be a shock risk if a line-to earth fault occurs?

The total loop impedance $Z_s = Z_e + R_1 + R_2$. We are given $Z_e = 0.27\,Ω$.

For a 6.0 mm² line conductor with a 2.5 mm² cpc, $R_1 + R_2$ is 10.49 mΩ/m. Hence, with a multiplier of 1.2 for 70°C PVC,

$$\text{total } R_1 + R_2 = \frac{18 \times 10.49 \times 1.2}{1000} = 0.23\,Ω$$

Therefore, $Z_s = 0.27 + 0.23 = 0.5\,Ω$. This is less than the 1.04 Ω maximum given in Table 41.2 for a 30 A BS 3036 fuse. Hence, the protection will disconnect the circuit in less than 0.4 s. In fact it will disconnect in less than 0.1 s; the determination of this time will be dealt with in Chapter 5.

Example 2.2

Consider, now, a more complex installation, and note how the procedure remains unchanged.

In this example, a three-phase motor is fed using 25 mm² single PVC conductors in trunking, the cpc being 2.5 mm². The circuit is protected by BS 88-3 45 A fuses in a distribution fuseboard. The distribution circuit or sub-main feeding this fuseboard comprises 70 mm² PVC singles in trunking with a 25 mm² cpc, the protection being by BS 88 160 A fuses. The external loop impedance Z_e has a measured value of 0.2 Ω. Will this circuit arrangement comply with the shock-risk constraints?

The formula $Z_s = Z_e + R_1 + R_2$ must be extended, as the $(R_1 + R_2)$ component comprises both the distribution and motor circuits; it therefore becomes:

$$Z_s = Z_e + (R_1 + R_2)_1 + (R_1 + R_2)_2$$

Distribution circuit $(R_1 + R_2)_1$

This comprises 30 m of 70 mm² line conductor and 30 m of 25 mm² cpc. Typical values for conductors over 35 mm² are shown in Table 2.2.

As an alternative we can use our knowledge of the relationship between conductor resistance and area, e.g. a 10 mm² conductor has approximately 10 times less resistance than a 1.0 mm² conductor:

10 mm² resistance = 1.83 mΩ/m
1.0 mm² resistance = 18.1 mΩ/m

Hence a 70 mm² conductor will have a resistance approximately half that of a 35 mm² conductor:

35 mm² resistance = 0.524 mΩ/m

$$\therefore 70 \text{ mm}^2 \text{ resistance} = \frac{0.524}{2} = 0.262 \text{ mΩ/m}$$

which compares well with the value given in Table 2.2.

25 mm² cpc resistance = 0.727 mΩ/m

so the distribution circuit

$$(R_1 + R_2)_1 = \frac{30 \times (0.262 + 0.727) \times 1.2}{1000} = 0.035 \, \Omega$$

Table 2.2   Resistance of Conductors

Area of Conductor (mm²)	Resistance (mΩ/m) Copper	Aluminium
50	0.387	0.641
70	0.263	0.443
95	0.193	0.320
120	0.153	0.253
185	0.0991	0.164
240	0.0754	0.125
300	0.0601	0.1

Hence $Z_s = Z_e + (R_1 + R_2)_1 = 0.2 + 0.035 = 0.235\,\Omega$, which is less than the Z_s maximum of $0.24\,\Omega$ quoted for a 160 A BS 88 fuse in Table 41.3 of the Regulations.

Motor circuit $(R_1 + R_2)_2$

Here we have 25 m of 25 mm^2 line conductor with 25 m of 2.5 mm^2 cpc. Hence:

$$(R_1 + R_2)_2 = \frac{25 \times (0.727 + 7.41) \times 1.2}{1000} = 0.24\,\Omega$$

$$\therefore \text{Total } Z_s = Z_e + (R_1 + R_2)_1 + (R_1 + R_2)_2 = 0.2 + 0.035 + 0.24 = 0.48\,\Omega$$

which is less than the Z_s maximum of $0.91\,\Omega$ quoted for a 45 A BS 88-3 fuse from Table 41.3 of the Regulations. Hence we have achieved compliance with the shock-risk constraints.

RESIDUAL CURRENT DEVICES

Additional protection can only be provided by RCDs with a rating $I_{\Delta n}$ not exceeding 30 mA and a tripping time not exceeding 40 ms at $5 \times I_{\Delta n}$.

The following list indicates the ratings and uses of RCDs detailed in BS 7671.

REQUIREMENTS FOR RCD PROTECTION

30 mA

- All socket outlets rated at not more than 20 A and for un-supervised general use
- Mobile equipment rated at not more than 32 A for use outdoors
- All circuits in a bath/shower room
- Preferred for all circuits in a TT system
- All cables installed less than 50 mm from the surface of a wall or partition unless they are adequately mechanically protected or have earthed screening or armouring, and also at any depth if the construction of the wall or partition includes metallic parts
- In zones 0, 1 and 2 of swimming pool locations

- All circuits in a location containing saunas, etc.
- Socket outlet final circuits not exceeding 32 A in agricultural locations
- Circuits supplying Class II equipment in restrictive conductive locations
- Each socket outlet in caravan parks and marinas and final circuit for houseboats
- All socket outlet circuits rated not more than 32 A for show stands, etc.
- All socket outlet circuits rated not more than 32 A for construction sites (where reduced low voltage, etc. is not used)
- All socket outlets supplying equipment outside mobile or transportable units
- All circuits in caravans
- All circuits in circuses, etc.
- A circuit supplying Class II heating equipment for floor and ceiling heating systems.

100 mA

- Socket outlets of rating exceeding 32 A in agricultural locations.

300 mA

- At the origin of a temporary supply to circuses, etc.
- Where there is a risk of fire due to storage of combustible materials
- All circuits (except socket outlets) in agricultural locations.

500 mA

- Any circuit supplying one or more socket outlets of rating exceeding 32 A, on a construction site.

We have seen the importance of the total earth loop impedance Z_s in the reduction of shock risk.

However, in some systems and especially TT, where the maximum values of Z_s given in Tables 41.2–41.4 of the Regulations may be hard to satisfy, an RCD may be used: its residual rating being determined from:

$$I_{\Delta n} \leq \frac{50}{Z_s}$$

In this case the maximum values of loop impedance for comparison are in BS 7671, Table 41.5.

Principle of operation of an RCD

Figure 2.13 illustrates the construction of an RCD. In a healthy circuit the same current passes through the line coil, the load, and back through the neutral coil. Hence the magnetic effects of line and neutral currents cancel out.

In a faulty circuit, either line to earth or neutral to earth, these currents are no longer equal. Therefore the out-of-balance current produces some residual magnetism in the core. As this magnetism is alternating, it links with the turns of the search coil, inducing an electro-motive force (EMF) in it. This EMF in turn drives a current through the trip coil, causing operation of the tripping mechanism.

It should be noted that a line-to-neutral fault will appear as a load, and hence the RCD will not operate for this fault.

A three-phase RCD works on the same out-of-balance principle; in this case the currents flowing in the three lines when they are all equal sum to zero, hence there is no resultant magnetism. Even if they are unequal, the out-of-balance current flows in the neutral which cancels out this out-of-balance current. Figure 2.14 shows the arrangement of a three-phase RCD, and Figure 2.15 shows how it can be connected for use on

FIGURE 2.13 Residual current device.

single-phase circuits. Operation of the test facility on an RCD simply indicates the correct functioning of the trip mechanism. It should be operated quarterly!

Nuisance tripping

Certain appliances such as cookers, water heaters and freezers tend to have, by the nature of their construction and use, some leakage currents to earth. These are quite normal, but could cause the operation of an RCD protecting an entire installation. This can be overcome by using split-load consumer units, where socket outlet circuits are protected by a 30 mA RCD, leaving all other circuits controlled by a normal mains

FIGURE 2.14 Three-phase RCD.

FIGURE 2.15 Connections for single-phase.

switch. Better still, especially in TT systems, is the use of a 100 mA RCD for protecting circuits other than socket outlets.

Modern developments in CB, RCD and consumer unit design now make it easy to protect any individual circuit with a combined CB/RCD (RCBO), making the use of split-load boards unnecessary.

An exception to the 30 mA RCD requirement for socket outlet circuits can be achieved by providing an indication that a particular socket outlet or outlets are not for general use, e.g. freezers, etc. This, of course, means the installation of a separate non-RCD protected circuit.

Supplementary bonding (IET Regulations Section 415.2)

In general the only Supplementary bonding required is for special locations such as bathrooms (not always needed – see Chapter 7), swimming pools, agricultural premises, etc. and where disconnection times cannot be met.

By now we should know why bonding is necessary; the next question, however, is to what extent bonding should be carried out. This is perhaps answered best by means of question and answer examples:

1. **Do I need to bond the hot and cold taps and a metal kitchen sink together? Surely they are all joined anyway?**
   Provided that main protective bonding conductors have been correctly installed there is no specific requirement in BS 7671 to do this.

2. **Do I have to bond radiators in a premises to, say, metal-clad switches or socket outlets etc.?**
   Supplementary bonding is only necessary when extraneous conductive parts are simultaneously accessible with exposed conductive parts and the disconnection time for the circuit concerned cannot be achieved. In these circumstances the bonding conductor should have a resistance $R \leq 50/I_a$, where I_a is the operating current of the protection.

3. **Do I need to bond metal window frames?**
   In general, no. Apart from the fact that most window frames will not introduce a potential from anywhere, the part of the window most

likely to be touched is the opening portion, to which it would not be practicable to bond. There may be a case for the bonding if the frames were fortuitously touching structural steel work.

4. **What about bonding in bathrooms?**

Refer to Chapter 7.

5. **What size of bonding conductors should I use?**

Main protective bonding conductors should be not less than half the size of the main earthing conductor, subject to a minimum of 6.0 mm^2 or, where PME (TN-C-S) conditions are present, a minimum of 10.0 mm^2. For example, most new domestic installations now have a 16.0 mm^2 earthing conductor, so all main bonding will be in 10.0 mm^2. Supplementary bonding conductors are subject to a minimum of 2.5 mm^2 if mechanically protected or 4.0 mm^2 if not. However, if these bonding conductors are connected to exposed conductive parts, they must be the same size as the cpc connected to the exposed conductive part, once again subject to the minimum sizes mentioned. It is sometimes difficult to protect a bonding conductor mechanically throughout its length, and especially at terminations, so it is perhaps better to use 4.0 mm^2 as the minimum size.

6. **Do I have to bond free-standing metal cabinets, screens, workbenches, etc.?**

No. These items will not introduce a potential into the equipotential zone from outside, and cannot therefore be regarded as extraneous conductive parts.

The Faraday cage

In one of his many experiments, Michael Faraday (1791–1867) placed an assistant in an open-sided cube which was then covered in a conducting material and insulated from the floor. When this cage arrangement was charged to a high voltage, the assistant found that he could move freely within it, touching any of the sides, with no adverse effects. Faraday had, in fact, created an equipotential zone, and of course in a correctly bonded installation, we live and/or work in Faraday cages!

Questions

1. What is the conventional potential of the earth?
2. What is the most common method of connection to earth?
3. Which point on the DNO's supply transformer is connected to earth?
4. How far from an earth rod does its resistance area extend?
5. How is a ground level voltage gradient for an earth rod avoided?
6. What is created by bonding?
7. List the three common DNO earthing systems.
8. In the formula $Z_s = Z_e + R1 + R2$, what is represented by each of the symbols?
9. What is the maximum disconnection time for a 20 A final circuit supplied from a TN system?
10. Which circuits in a caravan should be protected by a 30 mA RCD?
11. Why won't a line-neutral fault cause an RCD to trip?
12. How often should an RCD's test facility be operated?

Answers

1. Zero volts
2. Rod electrodes
3. The Neutral
4. Approximately 2.5 to 3 metres
5. By placing underground in a pit
6. An equipotential zone
7. TT, TN-S, TN-C-S
8. Z_s = total loop impedance; Z_e = external loop impedance; R1 = resistance of line conductor; R2 = resistance of protective conductor
9. 0.4 secs
10. All circuits
11. The RCD will see it as a load and not an out of balance fault current
12. Quarterly

Protection

☞ Relevant IET parts, chapters and sections: Part 4, Chapters 41, 42, 43, 44; Part 5, Chapter 53

Important terms/topics covered in this chapter:

- **Arm's reach** A zone of accessibility to touch, extending from any point on a surface where persons usually stand or move about, to the limits which a person can reach with his hand in any direction without assistance

- **Barrier** A part providing a defined degree of protection against contact with live parts, from any usual direction

- **Basic protection** Protection against electric shock under fault-free conditions

- **Circuit protective conductor** A protective conductor connecting exposed conductive parts of equipment to the main earthing terminal

- **Class II equipment** Equipment in which protection against electric shock does not rely on basic insulation only, but in which additional safety precautions such as supplementary insulation are provided. There is no provision for the connection of exposed metalwork of the equipment to a protective conductor, and no reliance upon precautions to be taken in the fixed wiring of the installation

- **Design current** The magnitude of the current intended to be carried by a circuit in normal service

- **Enclosure** A part providing an appropriate degree of protection of equipment against certain external influences and a defined degree of protection against contact with live parts from any direction

- **Exposed conductive part** A conductive part of equipment which can be touched and which is not a live part but which may become live under fault conditions

- **External influence** Any influence external to an electrical installation which affects the design and safe operation of that installation

- **Extraneous conductive part** A conductive part liable to introduce a potential, generally earth potential, and not forming part of the electrical installation

- **Fault current** A current resulting from a fault
- **Fault protection** Protection against electric shock under single fault conditions
- **Insulation** Suitable non-conductive material enclosing, surrounding or supporting a conductor
- **Isolation** Cutting off an electrical installation, a circuit or an item of equipment from every source of electrical energy
- **Live part** A conductor or conductive part intended to be energized in normal use, including a neutral conductor but, by convention, not a PEN conductor
- **Obstacle** A part preventing unintentional contact with live parts but not preventing deliberate contact
- **Overcurrent** A current exceeding the rated value. For conductors the rated value is the current-carrying capacity
- **Overload** An overcurrent occurring in a circuit which is electrically sound
- **Residual current device (RCD)** An electromechanical switching device or association of devices intended to cause the opening of the contacts when the residual current attains a given value under specified conditions
- **Short-circuit current** An overcurrent resulting from a fault of negligible impedance between live conductors having a difference of potential under normal operating conditions
- **Skilled person** A person with technical knowledge or sufficient experience to enable him to avoid the dangers which electricity may create

By the end of this chapter the reader should:

- be aware of the influences from which systems must be protected,
- be able to select equipment etc. with respect to IP and IK codes,
- know and understand the measures that can be taken to protect against shock,
- understand the requirements for protection against overcurrent,
- be able to select protective devices suitable for various applications,
- understand the relationship between 'let-through' energy and the thermal capacity of cables,
- understand the importance of discrimination,

- be aware of the need to protect against the effects of overvoltage and undervoltage.

WHAT IS PROTECTION?

The meaning of the word 'protection', as used in the electrical industry, is no different to that in everyday use. People protect themselves against personal or financial loss by means of insurance and from injury or discomfort by the use of the correct protective clothing. They further protect their property by the installation of security measures such as locks and/or alarm systems. In the same way, electrical systems need:

1. to be protected against mechanical damage, the effects of the environment and electrical overcurrents; and
2. to be installed in such a fashion that persons and/or livestock are protected from the dangers that such an electrical installation may create.

Let us now look at these protective measures in more detail.

Protection against mechanical damage

The word 'mechanical' is somewhat misleading in that most of us associate it with machinery of some sort. In fact, a serious electrical overcurrent left uninterrupted for too long can cause distortion of conductors and degradation of insulation; both of these effects are considered to be mechanical damage.

However, let us start by considering the ways of preventing mechanical damage by physical impact and the like.

Cable construction

A cable comprises one or more conductors each covered with an insulating material. This insulation provides protection from shock by contact with live parts and prevents the passage of leakage currents between conductors.

FIGURE 3.1   MI cable. On impact, all parts including the conductors are flattened, and a proportionate thickness of insulation remains between conductors, and conductors and sheath, without impairing the performance of the cable at normal working voltages.

Clearly, insulation is very important and in itself should be protected from damage. This may be achieved by covering the insulated conductors with a protective sheathing during manufacture, or by enclosing them in conduit or trunking at the installation stage.

The type of sheathing chosen and/or the installation method will depend on the environment in which the cable is to be installed. For example, metal conduit with thermoplastic (PVC) singles or mineral-insulated (MI) cable would be used in preference to PVC-sheathed cable clipped direct, in an industrial environment. Figure 3.1 shows the effect of physical impact on MI cable.

Protection against corrosion

Mechanical damage to cable sheaths and metalwork of wiring systems can occur through corrosion, and hence care must be taken to choose corrosion-resistant materials and to avoid contact between dissimilar metals in damp situations.

Protection against thermal effects

This is the subject of Chapter 42 of the IET Regulations. Basically, it requires common-sense decisions regarding the placing of fixed equipment, such that surrounding materials are not at risk from damage by heat.

Added to these requirements is the need to protect persons from burns by guarding parts of equipment liable to exceed temperatures listed in Table 42.1 of the Regulations.

Consumer units in dwellings now have to be manufactured from non-combustible material, e.g. steel, or be enclosed in a metal enclosure. This is in order to contain fire.

Polyvinyl chloride

PVC is a thermoplastic polymer widely used in electrical installation work for cable insulation, conduit and trunking. General-purpose PVC is manufactured to the British Standard BS 6746.

PVC in its raw state is a white powder; it is only after the addition of plasticizers and stabilizers that it acquires the form that we are familiar with.

Degradation

All PVC polymers are degraded or reduced in quality by heat and light. Special stabilizers added during manufacture help to retard this degradation at high temperatures. However, it is recommended that PVC-sheathed cables or thermoplastic fittings for luminaires (light fittings) should not be installed where the temperature is likely to rise above 60°C. Cables insulated with high-temperature PVC (up to 80°C) should be used for drops to lampholders and entries into batten-holders. PVC conduit and trunking should not be used in temperatures above 60°C.

Embrittlement and cracking

PVC exposed to low temperatures becomes brittle and will easily crack if stressed. Although both rigid and flexible, PVC used in cables and conduit can reach as low as 5°C without becoming brittle; it is recommended that general-purpose PVC-insulated cables should not be installed in areas where the temperature is likely to be consistently below 0°C, and that PVC-insulated cable should not be handled unless the ambient temperature is above 0°C and unless the cable temperature has been above 0°C for at least 24 hours.

Table 3.1 Examples of Classifications of External Influences

Environment	Utilization	Building
Water	Capability	Materials
AD6 Waves	**BA3** Handicapped	**CA1** Non-combustible

Where rigid PVC conduit is to be installed in areas where the ambient temperature is below −5°C but not lower than −25°C, type B conduit manufactured to BS 4607 should be used.

When PVC-insulated cables are installed in loft spaces insulated with polystyrene granules, contact between the two polymers can cause the plasticizer in the PVC to migrate to the granules. This causes the PVC to harden and, although there is no change in the electrical properties, the insulation may crack if disturbed.

External influences

Appendix 5 of the IET Regulations classifies external influences which may affect an installation. This classification is divided into three sections, the environment (A), how that environment is utilized (B) and construction of buildings (C). The nature of any influence within each section is also represented by a letter, and the level of influence by a number. Table 3.1 gives examples of the classification.

With external influences included on drawings and in specifications, installations and materials used can be designed accordingly.

Protection against ingress of solid objects, liquid and impact

In order to protect equipment from damage by foreign bodies, liquid or impact and also to prevent persons from coming into contact with live or moving parts, such equipment is housed inside enclosures or cable management systems such as conduit, trunking ducts, etc.

The degree of protection offered by such an enclosure is the subject of BS EN 60529 and BS EN 62262, commonly known as the IP and IK codes, parts of which are as shown in the accompanying tables. It will be seen from the IP table that, for instance, an enclosure to IP56 is dustproof and waterproof (Tables 3.2 and 3.3).

Table 3.2	IP Codes
First Numeral	**Mechanical Protection**
0	No protection of persons against contact with live or moving parts inside the enclosure. No protection of equipment against ingress of solid foreign bodies.
1	Protection against accidental or inadvertent contact with live or moving parts inside the enclosure by a large surface of the human body, for example a hand, not for protection against deliberate access to such parts. Protection against ingress of large solid foreign bodies.
2	Protection against contact with live or moving parts inside the enclosure by fingers. Protection against ingress of medium-sized solid foreign bodies.
3	Protection against contact with live or moving parts inside the enclosure by tools, wires or such objects of thickness greater than 2.5 mm. Protection against ingress of small foreign bodies.
4	Protection against contact with live or moving parts inside the enclosure by tools, wires or such objects of thickness greater than 1 mm. Protection against ingress of small foreign bodies.
5	Complete protection against contact with live or moving parts inside the enclosures. Protection against harmful deposits of dust. The ingress of dust is not totally prevented, but dust cannot enter in an amount sufficient to interfere with satisfactory operation of the equipment enclosed.
6	Complete protection against contact with live or moving parts inside the enclosures. Protection against ingress of dust.
Second Numeral	**Liquid Protection**
0	No protection.
1	Protection against drops of condensed water. Drops of condensed water falling on the enclosure shall have no effect.
2	Protection against drops of liquid. Drops of falling liquid shall have no harmful effect when the enclosure is tilted at any angle up to 15° from the vertical.
3	Protection against rain. Water falling in rain at an angle equal to or smaller than 60° with respect to the vertical shall have no harmful effect.
4	Protection against splashing. Liquid splashed from any direction shall have no harmful effect.
5	Protection against water jets. Water projected by a nozzle from any direction under stated conditions shall have no harmful effect.

(Continued)

Table 3.2	IP Codes – Cont'd
Second Numeral	**Liquid Protection**
6	Protection against conditions on ships' decks (deck with watertight equipment). Water from heavy seas shall not enter the enclosures under prescribed conditions.
7.	Protection against immersion in water. It must not be possible for water to enter the enclosure under stated conditions of pressure and time.
8.	Protection against indefinite immersion in water under specified pressure. It must not be possible for water to enter the enclosure.
X	Indicates no specified protections.

Table 3.3	IK Codes – Protection Against Mechanical Impact
Code	**Level of protection**
00	No protection.
01 to 05	Impact<1 joule
06 500 g 20 cm	Impact 1 joule
07 500 g 40 cm	Impact 2 joules
08 1.7 kg 29.5 cm	Impact 5 joules
09 5 kg 20 cm	Impact 10 joules
10 5 kg 40 cm	Impact 20 joules

The most commonly quoted IP codes in the Regulations are IPXXB and IP2X (the X denotes that no protection is specified, *not* that no protection exists).

Hence, IP2X means that an enclosure can withstand the ingress of medium-sized solid foreign bodies (12.5 mm diameter), and a jointed test finger, known affectionately as the British Standard finger! IPXXB denotes protection against the test finger only.

For accessible horizontal top surfaces of enclosures the IP codes are IPXXD and IP4X. This indicates protection against small foreign bodies and a 1 mm diameter test wire. IPXXD is the 1 mm diameter wire only.

IET Regulations Section 522 gives details of the types of equipment, cables, enclosure, etc. that may be selected for certain environmental conditions, e.g. an enclosure housing equipment in an AD8 environment (under water) would need to be to IPX8.

PROTECTION AGAINST ELECTRIC SHOCK (IET REGULATIONS CHAPTER 41)

There are two ways of receiving an electric shock: by contact with intentionally live parts, and by contact with conductive parts made live due to a fault. It is obvious that we need to provide protection against both of these conditions.

Basic protection (IET Regulations Sections 410 to 417)

Clearly, it is not satisfactory to have live parts accessible to touch by persons or livestock. The IET Regulations recommend five ways of minimizing this danger:

1. By covering the live part or parts with insulation which can only be removed by destruction, e.g. cable insulation.
2. By placing the live part or parts behind a barrier or inside an enclosure providing protection to at least IPXXB or IP2X. In most cases, during the life of an installation it becomes necessary to open an enclosure or remove a barrier. Under these circumstances, this action should only be possible by the use of a key or tool, e.g. by using a screwdriver to open a junction box. Alternatively, access

should only be gained after the supply to the live parts has been disconnected, e.g. by isolation on the front of a control panel where the cover cannot be removed until the isolator is in the 'off' position. An intermediate barrier of at least IP2X or IPXXB will give protection when an enclosure is opened: a good example of this is the barrier inside distribution fuseboards, preventing accidental contact with incoming live feeds.

3. By placing obstacles to prevent unintentional approach to or contact with live parts. This method must only be used where skilled persons are working.

4. By placing out of arm's reach: for example, the high level of the bare conductors of travelling cranes.

5. By using an RCD as additional protection. Whilst not permitted as the sole means of protection, this is considered to reduce the risk associated with contact with live parts, provided that one of the other methods just mentioned is applied, and that the RCD has a rated operating current $I_{\triangle n}$ of not more than 30 mA and an operating time not exceeding 40 ms at 5 times $I_{\triangle n}$, i.e. 150 mA.

Fault protection (IET Regulations Sections 410 to 417)

Protective earthing, protective equipotential bonding and automatic disconnection in the event of a fault have already been discussed in Chapter 2. The other methods are as follows.

Protection by automatic disconnection of supply (IET Regulations Section 411)

This measure is a combination of basic and fault protection.

Double or reinforced insulation

Often referred to as Class II equipment, this is typical of modern appliances where there is no provision for the connection of a cpc. This does not mean that there should be no exposed conductive parts and that the casing of equipment should be of an insulating material; it simply indicates that live parts are so well insulated that faults from live to conductive parts cannot occur.

Non-conducting location (IET Regulations Section 418)

This is basically an area in which the floor, walls and ceiling are all insulated. Within such an area there must be no protective conductors, and socket outlets will have no earthing connections.

It must not be possible simultaneously to touch two exposed conductive parts, or an exposed conductive part and an extraneous conductive part. This requirement clearly prevents shock current passing through a person in the event of an earth fault, and the insulated construction prevents shock current passing to earth.

Earth-free local equipotential bonding (IET Regulations Section 418)

This is, in essence, a Faraday cage, where all metals are bonded together but not to earth. Obviously great care must be taken when entering such a zone in order to avoid differences in potential between inside and outside.

The areas mentioned in this and the previous method are very uncommon. Where they do exist, they should be under constant supervision to ensure that no additions or alterations can lessen the protection intended.

Electrical separation (IET Regulations Section 418)

This method relies on a supply from a safety source such as an isolating transformer to BS EN 60742 which has no earth connection on the secondary side. In the event of a circuit that is supplied from a source developing a live fault to an exposed conductive part, there would be no path for shock current to flow: see Figure 3.2.

Once again, great care must be taken to maintain the integrity of this type of system, as an inadvertent connection to earth, or interconnection with other circuits, would render the protection useless.

Exemptions (IET Regulations 410.3.9)

As with most sets of rules and regulations, there are certain areas which are exempt from the requirements. These are listed quite clearly in IET Regulations 410.3.9, and there is no point in repeating them all here.

FIGURE 3.2   Electrical separation.

However, one example is the dispensing of the need to earth exposed conductive parts such as small fixings, screws and rivets, provided that they cannot be touched or gripped by a major part of the human body (not less than 50 mm by 50 mm), and that it is difficult to make and maintain an earth connection.

SELV OR PELV

This is simply extra low voltage (less than 50 V a.c.) derived from a safety source such as a Class II safety isolating transformer to BS EN 61558-2-6; or a motor generator which has the same degree of isolation as the transformer; or a battery or diesel generator; or an electronic device such as a signal generator.

Live or exposed conductive parts of separated extra low voltage (SELV) circuits should not be connected to earth, or protective conductors of other circuits, and SELV or PELV circuit conductors should ideally be kept separate from those of other circuits. If this is not possible, then the SELV conductors should be insulated to the highest voltage present.

Obviously, plugs and sockets of SELV or PELV circuits should not be interchangeable with those of other circuits.

SELV or PELV circuits supplying socket outlets are mainly used for hand lamps or soldering irons, for example, in schools and colleges. Perhaps

FIGURE 3.3 Reduced low voltage system.

a more common example of an SELV or PELV circuit is a domestic bell installation, where the transformer is to BS EN 60742. Note that bell wire is usually only suitable for 50–60 V, which means that it should not be run together with circuit cables of higher voltages.

Reduced low-voltage systems (IET Regulations Section 411.8)

The Health and Safety Executive accepts that a voltage of 65 V to earth, three-phase, or 55 V to earth, single-phase, will give protection against severe electric shock. They therefore recommend that portable tools used on construction sites, etc. be fed from a 110 V centre-tapped transformer. Figure 3.3 shows how 55 V is derived. Earth fault loop impedance values for these systems may be taken from Table 41.6 of the Regulations.

PROTECTION AGAINST OVERCURRENT (IET REGULATIONS CHAPTER 43 AND DEFINITIONS)

An overcurrent is a current greater than the rated current of a circuit. It may occur in two ways:

1. As an overload current; or
2. As a fault current, which may be subdivided into:
   (a) A short-circuit current; and
   (b) An earth fault current.

These conditions need to be protected against in order to avoid damage to circuit conductors and equipment. In practice, fuses and circuit breakers will fulfil both of these needs.

Overloads

Overloads are overcurrents occurring in healthy circuits. They may be caused, for example, by faulty appliances or by surges due to motors starting or by plugging in too many appliances in a socket outlet circuit.

Short circuits and earth faults

A short-circuit current is the current that will flow when a 'dead short' occurs between live conductors (line-to-neutral for single-phase; line-to-line for three-phase). Earth fault current flows when there is a short between a line conductor and earth. Prospective short-circuit current (PSCC) and prospective earth fault current (PEFC) are collectively known as prospective fault current (PFC). The term is usually used to signify the value of fault current at fuse or circuit breaker positions.

PFC is of great importance. However, before discussing it or any other overcurrent further, it is perhaps wise to refresh our memories with regard to fuses and circuit breakers and their characteristics.

Fuses and circuit breakers

As we all know, a fuse is the weak link in a circuit which will break when too much current flows, thus protecting the circuit conductors from damage.

There are many different types and sizes of fuse, all designed to perform a certain function. The IET Regulations refer to only four of these: BS 3036, BS 88-2, BS88-3 (supersedes BS1361) and BS 1362 fuses. It is perhaps sensible to include, at this point, circuit breakers to BS 3871 and BS EN 60898.

Breaking capacity of fuses and circuit breakers (IET Regulations Section 434)

When a fault occurs, the current may, for a fraction of a second, reach hundreds or even thousands of amperes. The protective device must be able to break and, in the case of circuit breakers, make such a current without damage to its surroundings by arcing, overheating or the scattering of hot particles.

Tables 3.4 and 3.5 indicate the performance of circuit breakers and the more commonly used British Standard fuse links.

Table 3.4 Breaking Capacities

Circuit Breakers	Breaking Capacity (kA)	
BS 3871 Types 1, 2, 3, etc.	1	(M1)
	1.5	(M1.5)
	3	(M3)
	4.5	(M4.5)
	6	(M6)
	9	(M9)
BS EN 60898 Types B, C, D	I_{cn} 1.5	I_{cs} 1.5
	3	3
	6	6
	10	7.5
	15	7.5
	25	10

I_{cn} is the rated ultimate breaking capacity. I_{cs} is the maximum breaking capacity operation after which the breaker may still be used without loss of performance.

Table 3.5 British Standards for Fuse Links

	Standard	Current Rating	Voltage Rating
1	BS 2950	Range 0.05–25 A	Range 1000 V (0.05 A) to 32 V (25 A) a.c. and d.c.
2	BS 646	1, 2, 3 and 5 A	Up to 250 V a.c. and d.c.
3	BS 1362 cartridge	1, 2, 3, 5, 7, 10 and 13 A	Up to 250 V a.c.
4	BS 88-3 HRC cut-out fuses	5, 15, 20, 30, 45 and 60 A	Up to 250 V a.c.
5	BS 88-2	Four ranges, 2–1200 A	Up to 660 V, but normally 250 or 415 V a.c. and 250 or 500 V d.c.
6	BS 2692	Main range from 5 to 200 A; 0.5 to 3 A for voltage trans-former protective fuses	Range from 2.2 to 132 kV
7	BS 3036 rewirable	5, 15, 20, 30, 45, 60, 100, 150 and 200 A	Up to 250 V to earth
8	BS 4265	500 mA to 6.3 A, 32 mA to 2 A	Up to 250 V a.c.

(Continued)

Table 3.5 British Standards for Fuse Links – Cont'd

Breaking Capacity	Notes
1. Two or three times current rating	Cartridge fuse links for telecommunication and light electrical apparatus. Very low breaking capacity
2. 1000 A	Cartridge fuse intended for fused plugs and adapters to BS 546: 'round-pin' plugs
3. 6000 A	Cartridge fuse primarily intended for BS 1363: 'flat-pin' plugs
4. 16 500 A, 33 000 A	Cartridge fuse intended for use in domestic consumer units. The dimensions prevent interchangeability of fuse links which are not of the same current rating
5. Ranges from 10 000 to 80 000 A in four a.c. and three d.c. categories	Part 1 of Standard gives performance and dimensions of cartridge fuse links, whilst Part 2 gives performance and requirements of fuse carriers and fuse bases designed to accommodate fuse links complying with Part 1
6. Ranges from 25 to 750 MVA (main range), 50 to 2500 MVA (VT fuses)	Fuses for a.c. power circuits above 660 V
7. Ranges from 1000 to 12 000 A	Semi-enclosed fuses (the element is a replacement wire) for a.c. and d.c. circuits
8. 1500 A (high breaking capacity), 35 A (low breaking capacity)	Miniature fuse links for protection of appliances of up to 250 V (metric standard)

Although all references to BS 3871 MCBs have been removed from BS 7671, they are still used and therefore worthy of mention.

Fuse and circuit breaker operation

Let us consider a protective device rated at, say, 10 A. This value of current can be carried indefinitely by the device, and is known as its nominal setting I_n. The value of the current which will cause operation of the device, I_2, will be larger than I_n, and will be dependent on the device's *fusing factor*. This is a figure which, when multiplied by the nominal setting I_n, will indicate the value of operating current I_2.

For fuses to BS 88-2, BS 88-3 and circuit breakers to BS 60898 this fusing factor is approximately 1.45; hence our 10 A device would not operate until the current reached $1.45 \times 10 = 14.5$ A.

The IET Regulations require coordination between conductors and protection when an overload occurs, such that:

1. The nominal setting of the device I_n is greater than or equal to the design current of the circuit I_b ($I_n \geq I_b$).
2. The nominal setting I_n is less than or equal to the lowest current-carrying capacity I_z of any of the circuit conductors ($I_n \leq I_z$).
3. The operating current of the device I_2 is less than or equal to $1.45I_z$ ($I_2 \leq 1.45I_z$).

So, for our 10 A device, if the cable is rated at 10 A then condition 2 is satisfied. Since the fusing factor is 1.45, condition 3 is also satisfied: $I_2 = I_n \times 1.45 = 10 \times 1.45$, which is also 1.45 times the 10 A cable rating.

The problem arises when a BS 3036 semi-enclosed rewirable fuse is used, as it may have a fusing factor of as much as 2. In order to comply with condition 3, I_n should be less than or equal to $0.725I_z$.

This figure is derived from $1.45/2 = 0.725$. For example, if a cable is rated at 10 A, then I_n for a BS 3036 should be $0.725 \times 10 = 7.25$ A. As the fusing factor is 2, the operating current $I_2 = 2 \times 7.25 = 14.5$, which conforms with condition 3, i.e. $I_2 \leq 1.45 \times 10 = 14.5$.

All of these foregoing requirements ensure that conductor insulation is undamaged when an overload occurs.

Under fault conditions it is the conductor itself that is susceptible to damage and must be protected. Figure 3.4 shows one half-cycle of short-circuit current if there were no protection. The RMS value ($0.7071 \times$ maximum value) is called the PFC. The cut-off point is where the fault current is interrupted and an arc is formed; the time t_1 taken to reach this point is called the pre-arcing time. After the current has been cut off, it falls to zero as the arc is being extinguished. The time t_2 is the total time taken to disconnect the fault.

During the time t_1, the protective device is allowing energy to pass through to the load side of the circuit. This energy is known as the pre-arcing let-through energy and is given by I^2t_1, where I is the fault current. The total let-through energy from start to disconnection of the fault is given by I^2t_2 (see Figure 3.5 and Table 3.6).

FIGURE 3.4 Let-through energy.

FIGURE 3.5 Let-through energy.

For faults of up to 5 s duration, the amount of heat energy that cable can withstand is given by k^2s^2, where s is the cross-sectional area of the conductor and k is a factor dependent on the conductor material. Hence, the let-through energy should not exceed k^2s^2, i.e. $I^2t \leq k^2s^2$. If we transpose this formula for t, we get $t \leq k^2s^2/I^2$, which is the maximum disconnection time in seconds.

Table 3.6 I^2t Characteristics: 2–800 A Fuse Links. Discrimination is Achieved if the Total I^2t of the Minor Fuse Does Not Exceed the Pre-Arcing I^2t of the Major Fuse

Rating (A)	I^2t Pre-Arcing	I^2t Total at 400 V
2	0.9	17
4	4	12
6	16	59
10	56	170
16	190	580
20	310	810
25	630	1700
32	1200	2800
40	2000	6000
50	3600	11 000
63	6500	14 000
80	13 000	36 000
100	24 000	66 000
125	34 000	120 000
160	80 000	260 000
200	140 000	400 000
250	230 000	560 000
315	360 000	920 000
350	550 000	1 300 000
400	800 000	2 300 000
450	700 000	1 400 000
500	900 000	1 800 000
630	2 200 000	4 500 000
700	2 500 000	5 000 000
800	4 300 000	10 000 000

Remember that these requirements refer to fault currents only. If, in fact, the protective device has been selected to protect against overloads and has a breaking capacity not less than the PFC at the point of installation, it will also protect against fault currents. However, if there is any doubt the formula should be used.

For example, in Figure 3.6, if I_n has been selected for overload protection, the questions to be asked are as follows:

1.	Is $I_n \geq I_b$?	Yes
2.	Is $I_n \leq I_z$?	Yes
3.	Is $I_2 \geq 1.45 I_z$?	Yes

Then, if the device has a rated breaking capacity not less than the PFC, it can be considered to give protection against fault current also.

When an installation is being designed, the PFC at every relevant point must be determined, by either calculation or measurement. The value will decrease as we move farther away from the intake position (resistance increases with length). Thus, if the breaking capacity of the lowest rated fuse in the installation is greater than the PFC at the origin of the supply, there is no need to determine the value except at the origin.

Discrimination (IET Regulation 536.2)

When we discriminate, we indicate our preference over other choices: this house rather than that house, for example. With protection we have to ensure that the correct device operates when there is a fault. Hence, a 13 A BS 1362 plug fuse should operate before the main circuit fuse. Logically, protection starts at the origin of an installation with a large device and progresses down the chain with smaller and smaller sizes.

BS 88 fuse (I_n) = 20 A, $I_2 = I_n \times 1.45 = 29$ A

Breaking capacity 1500 A

2.5 mm^2 conductors, rated at (I_z) 27 A

3 kW load
I_b = (3000/230)
= 13.04 A

Prospective fault current 800 A

FIGURE 3.6 Breaking capacity and PFC.

Simply because protective devices have different ratings, it cannot be assumed that discrimination is achieved. This is especially the case where a mixture of different types of device is used. However, as a general rule a 2:1 ratio with the lower-rated devices will be satisfactory. The table on page 57 shows how fuse links may be chosen to ensure discrimination.

Fuses will give discrimination if the figure in column 3 does not exceed the figure in column 2. Hence:

  a 2 A fuse will discriminate with a 4 A fuse
  a 4 A fuse will discriminate with a 6 A fuse
  a 6 A fuse will *not* discriminate with a 10 A fuse
  a 10 A fuse will discriminate with a 16 A fuse.

All other fuses will *not* discriminate with the next highest fuse, and in some cases several sizes higher are needed, e.g. a 250 A fuse will only discriminate with a 400 A fuse.

Position of protective devices (IET Regulations 433.2 and 434.2)

When there is a reduction in the current-carrying capacity of a conductor, a protective device is required. There are, however, some exceptions to this requirement; these are listed quite clearly in Sections 433 and 434 of the IET Regulations. As an example, protection is not needed in a ceiling rose where the cable size changes from $1.0\,mm^2$ to, say, $0.5\,mm^2$ for the lampholder flex. This is permitted as it is not expected that lamps will cause overloads.

PROTECTION AGAINST OVERVOLTAGE (IET REGULATIONS SECTION 443)

This chapter deals with the requirements of an electrical installation to withstand overvoltages caused by lightning or switching surges. Provided equipment, by virtue of its product specification, can withstand at least the values give in the tables, then no action is needed. However IT equipment is very susceptible and unless integral provision exists, surge protective devices may need to be installed.

PROTECTION AGAINST UNDERVOLTAGE (IET REGULATIONS SECTION 445)

From the point of view of danger in the event of a drop or loss of voltage, the protection should prevent automatic restarting of machinery, etc. In fact, such protection is an integral part of motor starters in the form of the control circuit.

Questions

1. What would be the minimum IP code for an item of equipment subject to splashes?
2. What IK code would indicate protection against an impact of 5 joules?
3. What is 'basic protection'?
4. What is 'fault protection'?
5. What system is recommended for the supply to portable tools on a construction site?
6. What is the fusing factor of a BS88 fuse?
7. What is the relationship between the breaking capacity of a protective device and prospective fault current?
8. What is total let-through energy?
9. What is the relationship between 'let-through' energy and the thermal capacity of a cable?
10. What is meant by discrimination between protective devices?

Answers

1. IP X4
2. IK 08
3. Protection against shock under fault free conditions
4. Protection against shock under single fault conditions
5. Reduced low voltage
6. 1.45
7. It must break, and in the case of a circuit breaker, make, the PFC at the point that it is installed without damage to surrounding equipment etc.
8. The electrical energy that a protective device lets through for the duration of a fault, and is given by I^2t
9. The let-through energy must not exceed the thermal capacity of the cable hence $I^2t \leq K^2S^2$
10. Ensuring that a minor protective device operates before a major device

Isolation Switching and Control

Important terms/topics covered in this chapter:

- **Emergency switching** Rapid cutting off of electrical energy to remove any hazard to persons, livestock or property which may occur unexpectedly
- **Isolation** Cutting off an electrical installation, a circuit or an item of equipment from every source of electrical energy
- **Mechanical maintenance** The replacement, refurbishment or cleaning of lamps and non-electrical parts of equipment, plant and machinery
- **Switch** A mechanical switching device capable of making, carrying and breaking current under normal circuit conditions, which may include specified overload conditions, and also of carrying, for a specified time, currents under specified abnormal conditions such as those of short circuit

By the end of this chapter the reader should:

- be aware of the reasons for isolation and switching,
- know the types of switching required in various types of installation,
- be able to select various general types of device to perform relevant switching activities.

ISOLATION AND SWITCHING (IET REGULATIONS SECTION 537)

All installations, whether they be the whole or part, must have a means of isolation and switching for various reasons. These are:

1. To remove possible dangers associated with the installation/operation/testing of electrical installations.
2. To provide a means of functional switching and control.

The IET Regulations make reference to:

1. **Switching off for mechanical maintenance** The devices for this function should be manually operated and preferably located in the main supply circuit.

2. **Emergency switching** The devices for this function should preferably be hand operated and be capable of interrupting the full load of the circuit concerned.

3. **Functional switching** This is simply switching an item on or off to control its function, e.g. a light switch.

4. **Firefighters' switches** Clearly for the function of isolation in the event of a fire. They should be coloured red and be installed no more than 2.75 m above the ground with the OFF position at the top.

The following chart (Table 4.1) shows type and uses of various devices used for isolation and switching.

Control

Motor control

This is usually part of the motor starter and most importantly must prevent automatic restarting after loss of supply and subsequent restoration, i.e. undervoltage protection.

Table 4.1   Selection of Generally Used Devices

Device	Isolation	Emergency	Function
Circuit breakers	Yes	Yes	Yes
RCDs	Yes	Yes	Yes
Isolating switches	Yes	Yes	Yes
Plugs and socket outlets	Yes	No	Yes
Ditto but over 32 A	Yes	No	No
Switched fused connection unit	Yes	Yes	Yes
Unswitched fused connection unit	Yes	No	No
Plug fuses	Yes	No	No
Cooker units	Yes	Yes	Yes

Auxiliary circuits (IET Regulations Section 557)

These are either a.c. or d.c. circuits used for the control, detection, measurements etc. of systems and equipment. The motor control circuit mentioned above is an auxiliary circuit, as is, for example, the smoke detector circuit for a fire alarm system.

Questions

1. What device is used for making, carrying and breaking current under normal circuit conditions?
2. Where, preferably, should a device for switching off for mechanical maintenance be located?
3. Where should the OFF position on a Fireman's switch be located?
4. What devices may be used for isolation and functional switching but not emergency switching?
5. What form of protection is achieved by the starter coil in a motor control circuit?

Answers

1. A switch
2. In the main supply circuit
3. At the top of the switch
4. Plugs and socket outlets
5. Undervoltage protection

Circuit Design

Important terms/topics covered in this chapter:

- **Ambient temperature** The temperature of the air or other medium where the equipment is to be used
- **Circuit protective conductor** A protective conductor connecting exposed conductive parts of equipment to the main earthing terminal
- **Current-carrying capacity** The maximum current which can be carried by a conductor under specified conditions without its steady state temperature exceeding a specified value
- **Design current** The magnitude of the current intended to be carried by a circuit in normal service
- **Earthing conductor** A protective conductor connecting a main earthing terminal of an installation to an earth electrode or other means of earthing
- **Fault current** An overcurrent resulting from a fault of negligible impedance between live conductors (short-circuit current) or between a line conductor and earth (earth fault current)
- **Overcurrent** A current exceeding the rated value. For conductors the rated value is the current-carrying capacity

By the end of this chapter the reader should:

- know the sequence of steps in a simple design procedure,
- be able to calculate single- and three-phase design currents,
- be able to select suitable protective devices,
- understand the importance and application of rating factors,
- be able to select appropriate cable sizes for current carrying capacity and voltage drop,
- understand the determination of shock risk and cable thermal constraints,
- be able to carry out simple design calculations.

DESIGN PROCEDURE

The requirements of IET Regulations make it clear that circuits must be designed and the design data made readily available. In fact, this has always been the case with previous editions of the Regulations, but it has not been so clearly indicated.

How then do we begin to design? Clearly, plunging into calculations of cable size is of little value unless the type of cable and its method of installation are known. This, in turn, will depend on the installation's environment. At the same time, we would need to know whether the supply was single- or three-phase, the type of earthing arrangements, and so on. Here then is our starting point and it is referred to in the Regulations, Chapter 3, as 'Assessment of general characteristics'.

Having ascertained all the necessary details, we can decide on an installation method, the type of cable, and how we will protect against electric shock and overcurrents. We would now be ready to begin the calculation part of the design procedure.

Basically there are eight stages in such a procedure. These are the same whatever the type of installation, be it a cooker circuit or a distribution cable feeding a distribution board in a factory. Here, then, are the eight basic steps in a simplified form:

1. Determine the design current I_b.
2. Select the rating of the protection I_n.
3. Select the relevant rating factors (CFs).
4. Divide I_n by the relevant CFs to give tabulated cable current-carrying capacity I_t.
5. Choose a cable size, to suit I_t.
6. Check the voltage drop.
7. Check for shock risk constraints.
8. Check for thermal constraints.

Let us now examine each stage in detail.

Add to this the requirement to select conduit and trunking sizes and we have a complete design.

DESIGN CURRENT

In many instances the design current I_b is quoted by the manufacturer, but there are times when it has to be calculated. In that case there are two formulae involved, one for single-phase and one for three-phase:

Single-phase:

$$I_b = \frac{P(\text{watts})}{V} (V \text{ usually } 230 \text{ V})$$

Three-phase:

$$I_b = \frac{P(\text{watts})}{\sqrt{3} \times V_L} (V_L \text{ usually } 400 \text{ V})$$

Current is in amperes, and power P in watts.

If an item of equipment has a power factor (PF) and/or has moving parts, efficiency (eff) will have to be taken into account.

Hence:

Single-phase:

$$I_b = \frac{P(\text{watts}) \times 100}{V \times \text{PF} \times \text{eff}}$$

Three-phase:

$$I_b = \frac{P \times 100}{\sqrt{3} \times V_L \times \text{PF} \times \text{eff}}$$

NOMINAL SETTING OF PROTECTION

Having determined I_b we must now select the nominal setting of the protection such that $I_n \geq I_b$. This value may be taken from IET Regulations, Tables 41.2, 41.3 or 41.4 or from manufacturers' charts. The choice of fuse or CB type is also important and may have to be changed if cable sizes or loop impedances are too high. These details will be discussed later.

Rating factors

When a cable carries its full load current it can become warm. This is no problem unless its temperature rises further due to other influences, in which case the insulation could be damaged by overheating. These other influences are: high ambient temperature; cables grouped together closely; uncleared overcurrents; and contact with thermal insulation.

For each of these conditions there is a rating factor (CF) which will respectively be called C_a, C_g, C_f and C_i, and which de-rates cable current-carrying capacity or conversely increases cable size.

Ambient temperature C_a

The cable ratings in the IET Regulations are based on an ambient temperature of 30°C, and hence it is only above this temperature that an adverse correction is needed. Table 4B1 of the Regulations gives factors for all types of insulation.

Grouping C_g

When cables are grouped together they impart heat to each other. Therefore, the more cables there are the more heat they will generate, thus increasing the temperature of each cable. Table 4C1 of the Regulations gives factors for such groups of cables or circuits. It should be noted that the figures given are for uniform groups of cables equally loaded, and hence correction may not necessarily be needed for cables grouped at the outlet of a domestic consumer unit, for example where there is a mixture of different sizes.

A typical situation where rating factors need to be applied would be in the calculation of cable sizes for a lighting system in a large factory. Here many cables of the same size and loading may be grouped together in trunking and could be expected to be fully loaded all at the same time.

Protection by BS 3036 fuse C_f

As we have already discussed in Chapter 3, because of the high fusing factor of BS 3036 fuses, the rating of the fuse I_n should be $\leq 0.725 I_z$.

Hence 0.725 is the rating factor to be used when BS 3036 fuses are used.

Thermal insulation C_i

With the modern trend towards energy saving and the installation of thermal insulation, there may be a need to de-rate cables to account for heat retention.

The values of cable current-carrying capacity given in Appendix 4 of the IET Regulations have been adjusted for situations when thermal insulation touches one side of a cable. However, if a cable is totally surrounded by thermal insulation for more than 0.5 m, a factor of 0.5 must be applied to the tabulated clipped direct ratings. For less than 0.5 m, de-rating factors (Table 52.2 of the Regulations) should be applied.

Application of rating factors

Some or all of the onerous conditions just outlined may affect a cable along its whole length or parts of it, but not all may affect it at the same time. So, consider the following:

1. If the cable in Figure 5.1 ran for the whole of its length, grouped with others of the same size in a high ambient temperature, and was totally surrounded with thermal insulation, it would seem logical to apply all the CFs, as they all affect the whole cable run. Certainly the factors for the BS 3036 fuse, grouping and thermal insulation should be used. However, it is doubtful if the ambient temperature will have any effect on the cable, as the thermal insulation, if it is efficient, will prevent heat reaching the cable. Hence, apply C_g, C_f and C_i.

FIGURE 5.1   Conditions along cable route.

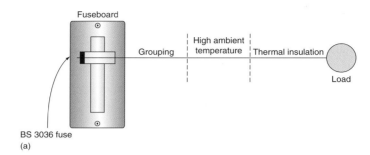

Fuseboard

Grouping | High ambient temperature | Thermal insulation

Load

BS 3036 fuse
(a)

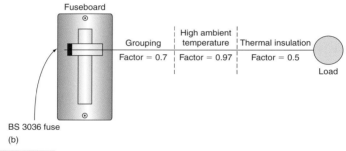

Fuseboard

Grouping | High ambient temperature | Thermal insulation
Factor = 0.7 | Factor = 0.97 | Factor = 0.5

Load

BS 3036 fuse
(b)

FIGURE 5.2   Conditions along cable route.

2. In Figure 5.2a the cable first runs grouped, then leaves the group and runs in high ambient temperature, and finally is enclosed in thermal insulation. We therefore have three different conditions, each affecting the cable in different areas. The BS 3036 fuse affects the whole cable run and therefore C_f must be used, but there is no need to apply all of the remaining factors as the worse one will automatically compensate for the others. The relevant factors are shown in Figure 5.2b; apply only $C_f = 0.725$ and $C_i = 0.5$. If protection was *not* by BS 3036 fuse, then apply only $C_i = 0.5$.

3. In Figure 5.3 a combination of cases 1 and 2 is considered. The effect of grouping and ambient temperature is $0.7 \times 0.97 = 0.69$. The factor for thermal insulation is still worse than this combination, and therefore C_i is the only one to be used.

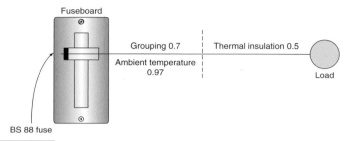

Fuseboard

Grouping 0.7

Thermal insulation 0.5

Ambient temperature 0.97

Load

BS 88 fuse

FIGURE 5.3   Conditions along cable route.

Having chosen the *relevant* rating factors, we now apply them to the nominal rating of the protection I_n as divisors in order to calculate the current-carrying capacity I_t of the cable.

Tabulated current-carrying capacity

The required formula for current-carrying capacity I_t is:

$$I_t \geq \frac{I_n}{\text{relevant CFs}}$$

In Figure 5.4 the current-carrying capacity is given by:

$$I_t \geq \frac{I_n}{C_f C_i} = \frac{30}{0.725 \times 0.5} = 82.75 \text{ A}$$

or, without the BS 3036 fuse:

$$I_t \geq \frac{30}{0.5} = 60 \text{ A}$$

In Figure 5.4, $C_a C_i = 0.97 \times 0.5 = 0.485$, which is worse than C_i (0.5) (Figure 5.5).

Hence:

$$I_t \geq \frac{I_n}{C_f C_a C_g} = \frac{30}{0.725 \times 0.485} = 85.3 \text{ A}$$

or, without the BS 3036 fuse:

$$I_t = \frac{30}{0.485} = 61.85 \text{ A}$$

Note: If the circuit is not subject to overload, I_n can be replaced by I_b so the formula becomes:

$$I_t \geq \frac{I_b}{\text{CFs}}$$

Choice of cable size

Having established the tabulated current-carrying capacity I_t of the cable to be used, it now remains to choose a cable to suit that value. The tables in Appendix 4 of the IET Regulations list all the cable sizes, current-carrying capacities and voltage drops of the various types of cable. For example, for PVC-insulated singles, single-phase, in conduit, having a current-carrying capacity of 45 A, the installation is by reference method B (Table

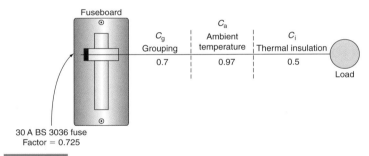

FIGURE 5.4 Conditions along cable route.

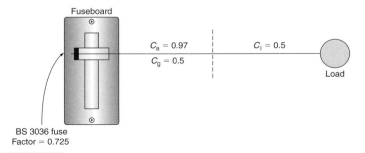

FIGURE 5.5 Conditions along cable route.

4A2), the cable table is 4D1A and the column is 4. Hence, the cable size is 10.0 mm² (column 1).

VOLTAGE DROP
(IET REGULATIONS 525 AND APPENDIX 12)

The resistance of a conductor increases as the length increases and/or the cross-sectional area decreases. Associated with an increased resistance is a drop in voltage, which means that a load at the end of a long thin cable will not have the full supply voltage available (Figure 5.6).

The IET Regulations require that the voltage drop V should not be so excessive that equipment does not function safely. They further indicate that the following percentages of the nominal voltage at the *origin* of the circuit will satisfy. This means that:

	LV Lighting (3%)	LV Power (5%)
230 V single-phase	6.9 V	11.5 V
400 V three-phase	12 V	20 V

For example, the voltage drop on a power circuit supplied from a 230 V source by a 16.0 mm² two-core copper cable 23 m long, clipped direct and carrying a design current of 33 A, will be:

$$V_c = \frac{mV \times I_b \times L}{1000} (mV; \text{ from Table 4D2B}) = \frac{2.8 \times 33 \times 23}{1000} = 2.125 \text{ V}$$

FIGURE 5.6   Voltage drop.

As we know that the maximum voltage drop in this instance (230 V) is 11.5 V, we can determine the maximum length by transposing the formula:

$$\text{Maximum length} = \frac{V_c \times 1000}{\text{mV} \times I_b} = \frac{11.5 \times 1000}{2.8 \times 23} = 178 \text{ m}$$

There are other constraints, however, which may not permit such a length.

SHOCK RISK
(IET REGULATIONS SECTION 411)

This topic has already been discussed in full in Chapter 2. To recap, however, the actual loop impedance Z_s should not exceed those values given in Tables 41.2, 41.3 and 41.4 of the IET Regulations. This ensures that circuits feeding final and distribution circuits will be disconnected, in the event of an earth fault, in the required time.

Remember $Z_s = Z_e + R_1 + R_2$.

THERMAL CONSTRAINTS
(IET REGULATIONS SECTION 543)

The IET Regulations require that we either select or check the size of a cpc against Table 54.7 of the Regulations, or calculate its size using an adiabatic equation.

Selection of cpc using Table 54.7

Table 54.7 of the Regulations simply tells us that:

1. For line conductors up to and including 16 mm², the cpc should be at least the same size.
2. For sizes between 16 mm² and 35 mm², the cpc should be at least 16 mm².
3. For sizes of line conductor over 35 mm², the cpc should be at least half this size.

This is all very well, but for large sizes of line conductor the cpc is also large and hence costly to supply and install. Also, composite cables such as the typical twin with cpc 6242Y type have cpcs smaller than the line conductor and hence do not comply with Table 54.7.

Calculation of cpc using an adiabatic equation

The adiabatic equation

$$s = \frac{\sqrt{I^2 t}}{k}$$

enables us to check on a selected size of cable, or on an actual size in a multicore cable. In order to apply the equation we need first to calculate the earth fault current from:

$$I = \frac{U_0}{Z_s}$$

where U_0 is the nominal line voltage to earth (usually 230 V) and Z_s is the actual earth fault loop impedance. Next we select a k factor from Tables 54.2 to 54.7 of the Regulations, and then determine the disconnection time t from the relevant curve.

For those unfamiliar with such curves, using them may appear a daunting task. A brief explanation may help to dispel any fears. Referring to any of the curves for fuses in Appendix 3 of the IET Regulations, we can see that the current scale goes from 1 A to 10 000 A, and the time scale from 0.01 s to 10 000 s. One can imagine the difficulty in drawing a scale between 1 A and 10 000 A in divisions of 1 A, and so a logarithmic scale is used. This cramps the large scale into a small area. All the subdivisions between the major divisions increase in equal amounts depending on the major division boundaries; for example, all the subdivisions between 100 and 1000 are in amounts of 100 (Figure 5.7).

Figures 5.8 and 5.9 give the IET Regulations time/current curves for BS 88 fuses. Referring to the appropriate curve for a 32 A fuse (Figure 5.9), we find that a fault current of 200 A will cause disconnection of the supply in 0.6 s.

FIGURE 5.7   Log–Log graph.

FIGURE 5.8   Log–Log divisions.

Where a value falls between two subdivisions, for example 150 A, an estimate of its position must be made. Remember that even if the scale is not visible, it would be cramped at one end; so 150 A would not fall half-way between 100 A and 200 A (Figure 5.8).

It will be noted in Appendix 3 of the Regulations that each set of curves is accompanied by a table which indicates the current that causes operation of the protective device for disconnection times of 0.1 s, 0.4 s and 5 s.

FIGURE 5.9 Time/current characteristics for fuses to BS 88 Part 2. Example for 32 A fuse superimposed.

The IET Regulations curves for CBs to BS EN 60898 type B and RCBOs are shown in Figure 5.11.

Having found a disconnection time, we can now apply the formula.

EXAMPLE OF USE OF THE ADIABATIC EQUATION

Suppose that in a design the protection was by 40 A BS 88-2 fuse; we had chosen a 4.0 mm² copper cpc running with our line conductor; and the loop impedance Z_s was 1.15 Ω. Would the chosen cpc size be large enough to withstand damage in the event of an earth fault? We have:

$$I = \frac{U_0}{Z_s} = \frac{230}{1.15} = 200 \text{ A}$$

FIGURE 5.10   Time/current characteristics for fuses to BS 88 Part 2. Example for 40 A fuse superimposed.

From the appropriate curve for the 40 A BS 88-2 fuse (Figure 5.10), we obtain a disconnection time t of 2 s. From Table 54.3 of the Regulations, $k = 115$. Therefore the minimum size of cpc is given by:

$$s = \frac{\sqrt{I^2 t}}{k} = \frac{\sqrt{200^2 \times 2}}{115} = 2.46 \text{ mm}^2$$

So our 4.0 mm² cpc is acceptable. Beware of thinking that the answer means that we could change the 4.0 mm² for a 2.5 mm². If we did, the loop impedance would be different and hence I and t would change; the answer for s would probably tell us to use a 4.0 mm².

In the example shown, 's' is merely a check on the actual size chosen.

Installation methods (IET Regulations Table 4.2)

Figures 5.12–5.18 illustrate some of the common methods of cable installation.

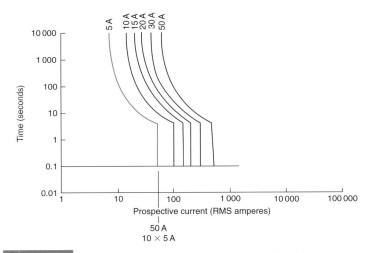

FIGURE 5.11   Time/current characteristics for type B CBs to BS EN 60898 and RCBOs. Example for 50 A fuse superimposed. For times less than 100 ms, the manufacturer should be consulted.

FIGURE 5.12   Method 100.

Having discussed each component of the design procedure, we can now put them all together to form a complete design.

AN EXAMPLE OF CIRCUIT DESIGN

A consumer lives in a bungalow with a detached garage and workshop, as shown in Figure 5.19 (see page 84). The building method is traditional brick and timber.

FIGURE 5.13   Method 101.

FIGURE 5.14   Method 102.

FIGURE 5.15   Method 103.

The mains intake position is at high level and comprises an 80 A BS 1361 230 V main fuse, an 80 A rated meter and a six-way 80 A consumer unit housing BS EN 60898 Type B CBs as follows:

FIGURE 5.16 Method A.

Ring circuit	32 A
Lighting circuit	6 A
Immersion heater circuit	16 A
Cooker circuit	32 A
Shower circuit	32 A
Spare way	–

The cooker is rated at 30 A, with no socket in the cooker unit. The main tails are 16.0 mm² double-insulated PVC, with a 6.0 mm² earthing conductor. There is no main protective bonding. The earthing system is TN-S, with an external loop impedance Z_e of 0.3 ohms. The prospective short-circuit current (PSCC) at the origin has been measured as 800 A. The roof space is insulated to full depth of the ceiling

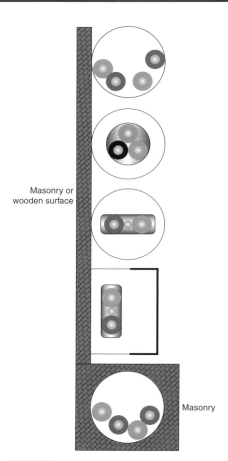

Masonry or
wooden surface

Masonry

FIGURE 5.17   Method B.

joists and the temperature in the roof space is not expected to exceed over 35°C.

The consumer wishes to convert the workshop into a pottery room and install an 8.6 kW/230 V electric kiln. The design procedure is as follows.

Masonry or
wooden
surface

Masonry

FIGURE 5.18 Method C.

ASSESSMENT OF GENERAL CHARACTERISTICS

The present maximum demand, applying diversity, is:

Ring	32 A
Lighting (66% of 6 A)	3.96 A
Immersion heater	16 A
Cooker (10 A + 30% of 20 A)	16 A
Shower	32 A
Total	**100 A**

8.6 kW, 230 V

Workshop   3 m

2.5 m

Garage   5 m

7 m

1 m

-------   Suggested
            cable route

8 m

Bungalow

FIGURE 5.19   Diagram for example on page 79.

Reference to the current rating tables in the IET Regulations will show that the existing main tails are too small and should be up-rated. So, the addition of another 8.6 kW of load is not possible with the present arrangement.

The current taken by the kiln is $8600/230 = 37.4$ A. Therefore, the new maximum demand is $100 + 37.4 = 137.4$ A.

Supply details are: single-phase 230 V, 50 Hz earthing; TN-S PSCC at origin (measured): 800 A.

Decisions must now be made as to the type of cable, the installation method and the type of protective device. As the existing arrangement is not satisfactory, the supply authority must be informed of the new maximum demand, as a larger main fuse and service cable may be required.

SIZING THE MAIN TAILS

1. The new load on the existing consumer unit will be 137.4 A. From the IET Regulations, the cable size is 25.0 mm².
2. The earthing conductor size, from the IET Regulations, will be 16.0 mm². The main protective bonding conductor size, from the IET Regulations, will be 10.0 mm².

For a domestic installation such as this, a PVC flat twin cable, clipped direct (avoiding any thermal insulation) through the loft space and the garage, etc., would be most appropriate.

SIZING THE KILN CIRCUIT CABLE

Design current I_b is:

$$I_b = \frac{P}{V} = \frac{8600}{230} = 37.4 \text{ A}$$

Rating and type of protection I_n

As we have seen, the requirement for the rating I_n is that $I_n \geq I_b$. Therefore, using the tables in the IET Regulations, I_n will be 40 A.

Rating factors:

C_a: 0.94
C_g: not applicable
C_f: 0.725 **only** if the fuse is BS 3036 (not applicable here)
C_i: 0.5 if the cable is totally surrounded in thermal insulation (not applicable here).

Tabulated current-carrying capacity of cable

$$I_t = \frac{I_n}{C_a} = \frac{40}{0.94} = 42.5 \text{ A}$$

Cable size based on tabulated current-carrying capacity

Table 4D5A IET Regulations gives a size of 6.0 mm² for this value of I_t (method C).

Check on voltage drop

The actual voltage drop is given by:

$$\frac{\mathrm{mV} \times I_b \times 1}{1000} = \frac{7.3 \times 37.4 \times 24.5}{1000} = 6.7 \text{ V (well below the maximum of } 11.5 \text{ V)}$$

This voltage drop, whilst not causing the kiln to work unsafely, may mean inefficiency, and it is perhaps better to use a 10.0 mm² cable.

For a 10.0 mm² cable, the voltage drop is checked as:

$$\frac{4.4 \times 37.4 \times 24.5}{1000} = 4.04 \text{ V}$$

Shock risk

The cpc associated with a 10.0 mm² twin 6242 Y cable is 4.0 mm². Hence, the total loop impedance will be:

$$Z_s = Z_e + \frac{(R_1 + R_2) \times L \times 1.2}{1000} = 0.3 + \frac{6.44 \times 24.5 \times 1.2}{1000} = 0.489 \ \Omega$$

Note

6.44 is the tabulated $(R_1 + R_2)$ value and the multiplier 1.2 takes account of the conductor resistance at its operating temperature.

It is likely that the chosen CB will be a type B.

Thermal constraints

We still need to check that the 4.0 mm² cpc is large enough to withstand damage under earth fault conditions. So, the fault current would be:

$$I = \frac{U_0}{Z_s} = \frac{230}{0.489} = 470 \text{ A}$$

The disconnection time t for this current for this type of protection (from the relevant curve in the IET Regulations) is as follows.

40 A CB type B = 0.1 s (the actual time is less than this but 0.1 s is the instantaneous time).

From the regulations, the factor for $k = 115$. We can now apply the adiabatic equation:

$$S = \frac{\sqrt{I^2 \times t}}{k} = \frac{\sqrt{470^2 \times 0.1}}{115} = 1.29 \text{ mm}^2$$

Hence, our 4.0 mm^2 cpc is of adequate size.

Summary

The kiln circuit would be protected by a 40 A BS EN 60898 type B CB and supplied from a spare way in the consumer unit. The main fuse would need to be up-rated to 100 A. The main tails would be changed to 25.0 mm^2. The earthing conductor would be changed to 16.0 mm^2.

Main protective bonding conductors would need to be 10.0 mm^2 twin with earth PVC cable.

Questions

1. What should be carried out prior to the commencement of design calculations?
2. What would be the design current of a 30 kW 400 V three-phase motor that is 85% efficient and has a p.f of 0.75?
3. What size of BS EN 60898 Type B circuit breaker would be required for a circuit with a design current of 68 A?
4. What is the rating factor for a group of 5 identical multicore cables installed, bunched in air to method C?
5. What de-rating factor would be applied to a cable passing horizontally through a thermally insulated stud wall 100 mm thick?
6. Determine the tabulated current carrying capacity I_t of a cable protected by a 15A BS 3036 fuse. The cable is not subjected to any adverse effects along its run.
7. Calculate the voltage drop for a 25 m length of 4.0 mm^2 flat twin with cpc cable if the protective device is rated at 32 A. Is this volt drop acceptable?
8. Is there a shock risk if a 20-metre long final circuit is protected by a 32 A BS 88-2 fuse, the conductor is 70° thermoplastic 6.0 mm2 with a 2.5 mm² cpc and the external loop impedance is 0.3Ω?
9. What current would flow in the event of an earth fault at the end of the circuit in question 8 above?
10. What minimum size of cpc would be required for a circuit if a fault current of 500 A disconnected in 0.2 seconds in a cable having a 'k' factor of 100?

Answers

1. An assessment of general characteristics
2. 68 A ($30,000 \times 100/\sqrt{3} \times 400 \times 85 \times 0.75$)
3. 80 A (table 41.3)
4. 0.6 (table 4C1)
5. 0.78 (table 52.2)
6. 20.7 A (15/0.725)
7. 8.8 V ($11 \times 32 \times 25/1000$). Yes, acceptable as max. is 11.5 V
8. No, actual is 0.55 Ω ($0.3 + (10.49 \times 20 \times 1.2)/1000$). Max. is 1.3 Ω
9. 418 A ($I = U_0/Z_s = 230/0.55$
10. 2.5mm^2 ($s = \sqrt{I^2 t}/k = \sqrt{500^2 \times 0.2}/100$)

Inspection and Testing

Important terms/topics covered in this chapter:

- **Earth electrode** A conductor or group of conductors in intimate contact with and providing an electrical connection with earth
- **Earth fault loop impedance** The impedance of the earth fault loop (line-to-earth loop) starting and ending at the point of earth fault
- **Residual current device (RCD)** An electromechanical switching device or association of devices intended to cause the opening of the contacts when the residual current attains a given value under specified conditions
- **Ring final circuit** A final circuit arranged in the form of a ring and connected to a single point of supply

By the end of this chapter the reader should:

- know the sequence of tests to be carried out on an installation,
- be aware of the need to zero or null instruments before conducting low resistance tests,
- know the general procedure for conducting each of the tests,
- be able to interpret test results.

TESTING SEQUENCE (PART 7)

Having designed our installation, selected the appropriate materials and equipment, and installed the system, it now remains to put it into service. However, before this happens, the installation must be tested and inspected to ensure that it complies, as far as is practicable, with the IET Regulations. Note the word 'practicable'; it would be unreasonable, for example, to expect the whole length of a circuit cable to be inspected for defects, as this may mean lifting floorboards, etc.

Part 6 of the IET Regulations gives details of testing and inspection requirements. Unfortunately, these requirements pre-suppose that the person carrying out the testing is in possession of all the design data, which is only likely to be the case on the larger commercial or industrial projects. It may be wise for the person who will eventually sign the test certificate to indicate that the test and inspection were carried out as far as was possible in the absence of any design or other information.

However, let us continue by examining the required procedures. The Regulations initially call for a visual inspection, but some items such as correct connection of conductors, etc. can be done during the actual testing. A preferred sequence of tests is recommended, where relevant, and is as follows:

1. Continuity of protective conductors
2. Continuity of ring final circuit conductors
3. Insulation resistance
4. Protection by SELV or PELV or electrical separation
5. Protection by barriers and enclosures provided during erection
6. Insulation of non-conducting floors and walls
7. Polarity
8. Earth electrode resistance
9. Earth fault loop impedance
10. Additional protection
11. Prospective fault current (PFC)
12. Check of phase sequence
13. Functional testing
14. Verification of voltage drop.

Not all of the tests may be relevant, of course. For example, in a domestic installation (TN-S or TN-C-S) only tests 1, 2, 3, 7, 9, 10, 11 and 13 would be needed.

The Regulations indicate quite clearly the tests required in Part 6. Let us then take a closer look at some of them in order to understand the reasoning behind them.

Continuity of protective conductors

All protective conductors, including main protective and supplementary bonding conductors, must be tested for continuity using a low-reading ohmmeter.

Table 6.1 Conductor Resistance at 20°C

Conductor Size (mm²)	Resistance (mΩ/m)
1.0	18.1
1.5	12.1
2.5	7.41
4.0	4.61
6.0	3.08
10.0	1.83
16.0	1.15
25.0	0.727
35.0	0.524

For main protective bonding conductors there is no single fixed value of resistance above which the conductor would be deemed unsuitable. Each measured value, if indeed it is measurable for very short lengths, should be compared with the relevant value for a particular conductor length and size. Such values are shown in Table 6.1.

Where a supplementary protective bonding conductor has been installed between *simultaneously accessible* exposed and extraneous conductive parts, because circuit disconnection times cannot be met, then the resistance R of the conductor must be equal to or less than $50/I_a$. So:

$$R \leq \frac{50}{I_a}\ \Omega$$

where 50 is the voltage, above which exposed metalwork should not rise, and I_a is the minimum current, causing operation of the circuit protective device within 5 s.

For example, suppose a 45 A BS 3036 fuse protects a cooker circuit. The disconnection time for the circuit cannot be met, and so a supplementary bonding conductor has been installed between the cooker case and the adjacent metal sink. The resistance R of that conductor should not be greater than $50/I_a$, which in this case is 145 A (IET Regulations). So:

$$\frac{50}{145} = 0.34\ \Omega$$

How, then, do we conduct a test to establish continuity of main or supplementary bonding conductors? Quite simple really: just connect the leads from the continuity tester to the ends of the bonding conductor (Figure 6.1). One end should be disconnected from its bonding clamp, otherwise any measurement may include the resistance of parallel paths of other earthed metalwork. Remember to zero or null the instrument first or, if this facility is not available, record the resistance of the test leads so that this value can be subtracted from the test reading.

> **Important Note**
>
> If the installation is in operation, then never disconnect main bonding conductors unless the supply can be isolated. Without isolation, persons and livestock are at risk of electric shock.

The continuity of circuit protective conductors may be established in the same way, but a second method is preferred, as the results of this second test indicate the value of $(R_1 + R_2)$ for the circuit in question.

FIGURE 6.1   Continuity of main protective bonding conductor.

The test is conducted in the following manner:

1. Temporarily link together the line conductor and cpc of the circuit concerned in the distribution board or consumer unit.
2. Test between line and cpc at each outlet in the circuit. A reading indicates continuity.
3. Record the test result obtained at the furthest point in the circuit. This value is $(R_1 + R_2)$ for the circuit.

Figure 6.2 illustrates the above method.

There may be some difficulty in determining the $(R_1 + R_2)$ values of circuits in installations that comprise steel conduit and trunking, and/or SWA and mims cables because of the parallel earth paths that are likely to exist. In these cases, continuity tests may have to be carried out at the installation stage before accessories are connected or terminations made off as well as after completion.

Continuity of ring final circuit conductors

There are two main reasons for conducting this test:

1. To establish that interconnections in the ring do not exist.
2. To ensure that the circuit conductors are continuous, and indicate the value of $(R_1 + R_2)$ for the ring.

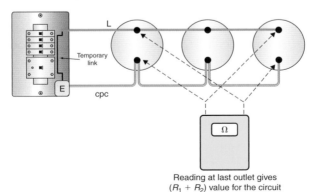

Reading at last outlet gives
$(R_1 + R_2)$ value for the circuit

FIGURE 6.2 cpc continuity.

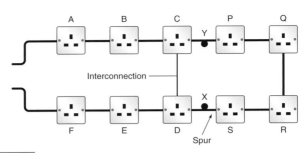

FIGURE 6.3 Ring circuit with interconnection.

What then are interconnections in a ring circuit, and why is it important to locate them? Figure 6.3 shows a ring final circuit with an interconnection.

The most likely cause of the situation shown in Figure 6.3 is where a DIY enthusiast has added sockets P, Q, R and S to an existing ring A, B, C, D, E and F. In itself there is nothing wrong with this. The problem arises if a break occurs at, say, point Y, or the terminations fail in socket C or P. Then there would be four sockets all fed from the point X which would then become a spur. So, how do we identify such a situation with or without breaks at point Y? A simple resistance test between the ends of the line, neutral or circuit protective conductors will only indicate that a circuit exists, whether there are interconnections or not. The following test method is based on the theory that the resistance measured across any diameter of a perfect circle of conductor will always be the same value (Figure 6.4).

The perfect circle of conductor is achieved by cross-connecting the line and neutral loops of the ring (Figure 6.5).

The test procedure is as follows:

1. *Identify the opposite legs of the ring.* This is quite easy with sheathed cables, but with singles, each conductor will have to be identified, probably by taking resistance measurements between each one and the closest socket outlet. This will give three high readings and three low readings, thus establishing the opposite legs.

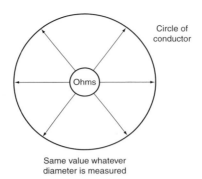

FIGURE 6.4 Resistance across a circle of conductor

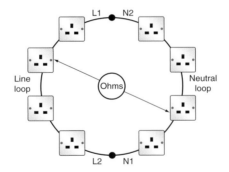

FIGURE 6.5 Circle formed by cross-connection.

2. *Take a resistance measurement between the ends of each conductor loop. Record this value.*
3. *Cross-connect the ends of the line and neutral loops (see Figure 6.6).*
4. *Measure between line and neutral at each socket on the ring.* The readings obtained should be, for a perfect ring, substantially the same.

If an interconnection existed such as shown in Figure 6.3 then sockets A to F would all have similar readings, and those beyond the interconnection would have gradually increasing values to approximately the

FIGURE 6.6 L and N cross-connection.

midpoint of the ring, then decreasing values back towards the interconnection. If a break had occurred at point Y then the readings from socket S would increase to a maximum at socket P. One or two high readings are likely to indicate either loose connections or spurs. A null reading, i.e. an open circuit indication, is probably a reverse polarity, either line–cpc or neutral–cpc reversal. These faults would clearly be rectified and the test at the suspect socket(s) repeated.

5. *Repeat the above procedure, but in this case cross-connect the line and cpc loops*. In this instance, if the cable is of the flat twin type, the readings at each socket will very slightly increase and then decrease around the ring. This difference, due to the line and cpc being different sizes, will not be significant enough to cause any concern. The measured value is very important it is $(R_1 + R_2)$ for the ring.

As before, loose connections, spurs and, in this case, L–N cross-polarity will be picked up.

The details that follow are typical approximate ohmic values for a healthy 70 m ring final circuit wired in $2.5\,mm^2/1.5\,mm^2$ flat twin and cpc cable:

Initial Measurements	L1–L2	N1–N2	cpc1–cpc2
Reading at each socket	0.26 Ω	0.26 Ω	between 0.32 Ω and 0.34 Ω
For spurs, each metre in length will add the following resistance to the above values	0.015 Ω	0.015 Ω	0.02 Ω

Conductors

Sheathing

Resistance of installation

FIGURE 6.7 Cable insulation resistance.

Insulation resistance

This is probably the most used and yet most abused test of them all. Affectionately known as 'meggering', an insulation resistance test is performed in order to ensure that the insulation of conductors, accessories and equipment is in a healthy condition, and will prevent dangerous leakage currents between conductors and between conductors and earth. It also indicates whether any short circuits exist.

Insulation resistance is the resistance measured between conductors and is made up of countless millions of resistances in parallel (Figure 6.7).

The more resistances there are in parallel, the lower the overall resistance, and in consequence, the longer a cable the lower the insulation resistance. Add to this the fact that almost all installation circuits are also wired in parallel, and it becomes apparent that tests on large installations may give, if measured as a whole, pessimistically low values, even if there are no faults. Under these circumstances, it is usual to break down such large installations into smaller sections, floor by floor, distribution circuit by distribution circuit, etc. This also helps, in the case of periodic testing, to minimize disruption.

The test procedure, then, is as follows:

1. Disconnect all items of equipment such as capacitors and indicator lamps as these are likely to give misleading results. Remove any items of equipment likely to be damaged by the test, such as dimmer switches, electronic timers, etc. Remove all lamps and accessories and disconnect fluorescent and discharge fittings. Ensure that the installation is disconnected from the supply, all fuses are in place, and CBs and switches are in the on position. In some instances it may be impracticable to remove lamps, etc. and in this case the local switch controlling such equipment may be left in the off position.

2. Join together all live conductors of the supply and test between this join and earth. Alternatively, test between each live conductor and earth in turn.

3. Test between line and neutral. For three-phase systems, join together all lines and test between this join and neutral. Then test between each of the lines. Alternatively, test between each of the live conductors in turn. Installations incorporating two-way lighting systems should be tested twice with the two-way switches in alternative positions.

Table 6.2 gives the test voltages and minimum values of insulation resistance for ELV and LV systems.

Table 6.2   Insulation Resistance Test Values

System	Test Voltage	Minimum Insulation Resistance
SELV and PELV	250 V d.c.	0.5 MΩ
LV up to 500 V	500 V d.c.	1.0 MΩ
Over 500 V	1000 V d.c.	1.0 MΩ

If a value of less than 2 MΩ is recorded it may indicate a situation where a fault is developing, but as yet still complies with the minimum permissible value. In this case, each circuit should be tested separately to identify any that are suspect.

Polarity

This simple test, often overlooked, is just as important as all the others, and many serious injuries and electrocutions could have been prevented if only polarity checks had been carried out.

The requirements are:

- all fuses and single pole switches are in the line conductor
- the centre contact of an Edison screw type lampholder is connected to the line conductor
- all socket outlets and similar accessories are correctly wired.

Although polarity is towards the end of the recommended test sequence, it would seem sensible, on lighting circuits for example, to conduct this test at the same time as that for continuity of cpcs (Figure 6.8).

As discussed earlier, polarity on ring final circuit conductors is achieved simply by conducting the ring circuit test. For radial socket outlet circuits, however, this is a little more difficult. The continuity of the cpc will have already been proved by linking line and cpc and measuring between the same terminals at each socket. Whilst a

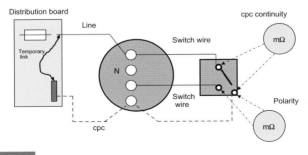

FIGURE 6.8 Lighting polarity.

line–cpc reversal would not have shown, a line–neutral reversal would, as there would have been no reading registered at the socket in question. This would have been remedied, and so only line–cpc reversals need to be checked. This can be done by linking together line and neutral at the origin and testing between the same terminals at each socket. A line–cpc reversal will result in no reading at the socket in question.

Earth electrode resistance

As we know, in many rural areas, the supply system is TT and hence reliance is placed on the general mass of earth for a return path under earth fault conditions and the connection to earth is made by an electrode, usually of the rod type.

In order to determine the earth electrode resistance an earth electrode resistance tester and auxiliary electrodes are used. If the installation is protected by an RCD a standard external impedance Z_e test is conducted with an earth fault loop impedance tester connected between the incoming line terminal and the top of the electrode.

Earth fault loop impedance

Overcurrent protective devices must, under earth fault conditions, disconnect fast enough to reduce the risk of electric shock. This is achieved if the actual value of the earth fault loop impedance does not exceed the tabulated maximum values given in BS 7671. The purpose of the test, therefore, is to determine the actual value of the loop impedance Z_s, for comparison with those maximum values, and it is conducted as follows:

1. Ensure that all main protective bonding is in place.
2. Connect the test instrument either by its BS 1363 plug, or the 'flying leads', to the line, neutral and earth terminals at the remote end of the circuit under test. (If a neutral is not available, connect the neutral probe to earth.)
3. Press to test and record the value indicated.

It must be understood that this instrument reading is *not valid for direct comparison with the tabulated maximum values*, as account must be taken of the ambient temperature at the time of test, and the maximum conductor operating temperature, both of which will have an effect on conductor resistance. Hence, the $(R_1 + R_2)$ is likely to be greater at the time of fault than at the time of test.

So, our measured value of Z_s must be corrected using correction factors and applying them in a formula.

Clearly this method of correcting Z_s is time-consuming and unlikely to be commonly used. Hence, a rule of thumb method may be applied which simply requires that the measured value of Z_s does not exceed 0.8 of the appropriate tabulated value. Table 6.3 gives the 0.8 values of tabulated loop impedance for direct comparison with measured values.

In effect, a loop impedance test places a line/earth fault on the installation, and if an RCD is present it may not be possible to conduct the test, as the device will trip out each time the loop impedance tester button is pressed. Unless the instrument is of a type that has a built-in guard against such tripping, the value of Z_s will have to be determined from measured values of Z_e and $(R_1 + R_2)$.

Important Note

Never short out an RCD in order to conduct this test.

As a loop impedance test creates a high earth fault current, albeit for a short space of time, some lower rated CBs may operate, resulting in the same situation as with an RCD, and Z_s will have to be calculated. It is not really good practice to temporarily replace the CB with one of a higher rating.

External loop impedance Z_e

The value of Z_e is measured at the intake position on the supply side and with all main protective bonding disconnected. Unless the installation can be isolated from the supply, this test should not be carried out, as a

Table 6.3 Corrected Maximum Z_s Values for Comparison with Measured Values

PROTECTION RATING

PROTECTIVE DEVICE		5	6	10	15	16	20	25	30	32	40	45	50	60	63	80	100	125	160	200
BS EN 60898 and 61009 type **B**	0.4s & 5s		5.82	3.5		2.19	1.75	1.4		1.09	0.87		0.7		0.56	0.5	0.35	0.29		
BS EN 60898 and 61009 type **C**	0.4s & 5s		2.9	1.75		1.09	0.87	0.71		0.54	0.44		0.35		0.28	0.2	0.17	0.14		
BS EN 60898 and 61009 type **D**	0.4s		1.46	0.87		0.54	0.44	0.35		0.28	0.22		0.17		0.13	0.1	0.08	0.07		
	5s		2.91	1.75		1.09	0.87			0.55	0.44		0.35		0.27	0.22	0.17	0.14		
BS 3036 Semi enclosed	0.4s	7.22			1.94	1.34			0.83											
	5s	13.45			4.1	2.9			2			1.2		0.85			0.4			
		3A*			13A*															
BS 1362 Cartridge	0.4s	12.5			1.8															
	5s	17.63			2.9															

BS 88-2. Bolted type (E) and Clip-in type (G)

0.4s		6.24	3.71		1.95	1.35	1.03		0.79						
5s		9.73	5.46		3.17	2.24	1.75	1.4	1.03	0.79	0.64	0.4	0.35	0.26	0.21	0.2

BS 88-3. Cartridge type (C)

0.4s		7.94			1.84	1.55		0.73				
5s		11.62			2.14	2.57		1.24	0.79	0.55	0.4	0.3

(Continued)

Table 6.3 Corrected Maximum Z_s Values for Comparison with Measured Values – Cont'd

OBSOLETE FUSES & CIRCUIT BREAKERS

MAXIMUM CORRECTED Z_s VALUES

PROTECTION RATING

		5	6	10	15	16	20	25	30	32	40	45	50	60	63	80	100	125	160	200
BS 88 – 2.2 & 88-6	0.4s		6.48	3.89		2.05	1.35	1.09		0.79										
	5s		10.26	5.64		3.16	2.2	1.75		1.4	1.03		0.79		0.64	0.5	0.32	0.25	0.19	0.2
BS 1361 Cartridge	0.4s	7.94			2.45		1.29		0.87											
	5s	12.46			3.8		2.13		1.4			0.75		0.53		0.4	0.28			
BS 3871 type 1	0.4s & 5s	8.74	7.22	4.37	2.9	2.73	2.19	1.75	1.45	1.37	1.09	0.97	0.87		0.69					
BS 3871 type 2	0.4s & 5s	5	4.15	2.49	1.66	1.56	1.24	1	0.83	0.78	0.64	0.55	0.49		0.42					
BS 3871 type 3	0.4s & 5s	3.5	2.85	1.75	1.16	1.09	0.87	0.7	0.58	0.54	0.44	0.39	0.35		0.28					

Table 6.4   RCD Tripping Times

RCD Type	Half Rated	Full Trip Current
BS 4293 sockets	No trip	Less than 200 ms
BS 4293 with time delay	No trip	½ time delay 1200 ms
BS EN 61009 or BS EN 61009 RCBO	No trip	300 ms
As above, but type S with time delay	No trip	130–500 ms

potential shock risk will exist with the supply on and the main protective bonding disconnected.

Additional protection RCD/RCBO operation

Where RCDs/RCBOs are fitted, it is essential that they operate within set parameters. The RCD testers used are designed to do just this, and the basic tests required are as follows:

1. Set the test instrument to the rating of the RCD.
2. Set the test instrument to half-rated trip.
3. Operate the instrument and the RCD should not trip.
4. Set the instrument to deliver the full rated tripping current of the RCD, $I_{\Delta n}$.
5. Operate the instrument and the RCD should trip out in the required time.
6. For RCDs rated at 30 mA or less set the instrument to deliver 5 times the rated tripping current of the RCD, $5I_{\Delta n}$.
7. Operate the instrument and the RCD should trip out in 40 ms.

Table 6.4 gives further details.

Prospective fault current

Prospective fault current (PFC) has to be determined at the origin of the installation. This is achieved by enquiry, calculation or measurement.

Check of phase sequence

Where multi-phase systems are used there is a high possibility that phase sequence will need to be checked.

This is done with the use of a phase rotation indicator, which, simplistically, is a small three-phase motor.

Functional testing

All RCDs have a built-in test facility in the form of a test button. Operating this test facility creates an artificial out-of-balance condition that causes the device to trip. This only checks the mechanics of the tripping operation; it is not a substitute for the tests just discussed.

All other items of equipment such as switchgear, control gear, interlocks, etc. must be checked to ensure that they are correctly mounted and adjusted and that they function correctly.

Verification of voltage drop

Where required the voltage drop to the furthest point of a circuit should be determined. This is not usually needed for initial verification.

Questions

1. Why should a test for continuity of a protective bonding conductor require the disconnection of that conductor?
2. What precaution should be taken before conducting a continuity test on a protective bonding conductor?
3. What is the significance of the reading obtained at the end of a circuit when conducting a cpc continuity test?
4. Apart from identifying interconnections and establishing a value for $(R_1 + R_2)$, what other test is automatically conducted when carrying out a ring final circuit continuity test?
5. For an insulation resistance test on a 600 V discharge lighting circuit, state the required test voltage and minimum value of insulation resistance.
6. What instrument is used for conducting a polarity test?
7. Why is a 0.8 factor used when comparing measured values of Z_s with maximum tabulated values?
8. Why must the earthing conductor of an installation be disconnected when conducting a test for external loop impedance?
9. What are the test requirements for a 20 mA RCD used for additional protection?
10. What is the purpose of the test function on an RCD?

Answers

1. To avoid parallel paths
2. Isolate the supply
3. $(R_1 + R_2)$
4. Polarity
5. 1000 V, 1 MΩ
6. Low resistance ohmmeter
7. to compensate for ambient and conductor temperature
8. To avoid parallel paths
9. Must disconnect in 40 ms at $5 \times I_{\Delta n}$ (100 mA)
10. To confirm that its mechanism is functioning correctly

Special Locations
IET Regulations Part 7

INTRODUCTION

The bulk of BS 7671 relates to typical, single- and three-phase, installations. There are, however, some special installations or locations that have particular requirements. Such locations may present the user/occupant with an increased risk of death or injuries from electric shock.

BS 7671 categorizes these special locations in Part 7 and they comprise the following:

Section 701	Bathrooms, shower rooms, etc.
Section 702	Swimming pools and other basins
Section 703	Rooms containing sauna heaters
Section 704	Construction and demolition sites
Section 705	Agricultural and horticultural premises
Section 706	Conducting locations with restrictive movement
Section 708	Caravan/camping parks
Section 709	Marinas and similar locations
Section 710	Medical locations
Section 711	Exhibitions, shows and stands
Section 712	Solar photovoltaic (PV) power supply systems
Section 714	Outdoor lighting installations
Section 715	Extra-low voltage installations
Section 722	Electric vehicle charging installations
Section 717	Mobile or transportable units
Section 721	Caravans and motor caravans
Section 729	Operating and maintenance gangways
Section 740	Amusement devices, fairgrounds, circuses, etc.
Section 753	Floor and ceiling heating systems

Let us now briefly investigate the main requirements for each of these special locations.

17th Edition IET Wiring Regulations: Explained and Illustrated. 978-1-138-84917-4.
© Brian Scaddan. Published by Taylor & Francis. All rights reserved.

BS 7671 SECTION 701: BATHROOMS, ETC.

This section deals with rooms containing baths, shower basins or areas where showers exist but with tiled floors, for example leisure/recreational centres, sports complexes, etc.

Each of these locations is divided into zones 0, 1 and 2, which give an indication of their extent and the equipment/wiring, etc. that can be installed in order to reduce the risk of electric shock.

So! Out with the tape measure, only to find that in a one-bedroom flat, there may be no zone 2. How can you conform to BS 7671?

The stark answer (mine) is that you may not be able to conform exactly. You do the very best you can in each particular circumstance to ensure safety. Let us not forget that the requirements of BS 7671 are based on reasonableness.

Zone 0

This is the interior of the bath tub or shower basin or, in the case of a shower area without a tray, it is the space having a depth of 100 mm above the floor out to a radius of 600 mm from a fixed water outlet or 1200 mm radius for an extended zone 1 (Figure 7.1).

Note

- Only SELV (12 V) or ripple-free d.c. may be used as a measure against electric shock, the safety source being outside zones 0, 1 and 2.
- Other than current using equipment specifically designed for use in this zone, **no** switchgear or accessories are permitted.
- Equipment designed for use in this zone must be to at least IPX7.
- Only wiring associated with equipment in this zone may be installed.

Zone 1

This extends above zone 0 around the perimeter of the bath or shower basin to 2.25 m above the floor level, and includes any space below the bath or basin that is accessible without the use of a key or tool. For showers without

FIGURE 7.1 Zone 0.

FIGURE 7.2 Zones 0 and 1.

basins, zone 1 extends out to a radius of 600 mm from a fixed water outlet or 1200 mm radius for an extended zone 1. (Figures 7.2 and 7.3).

Note

- Other than switches and controls of equipment specifically designed for use in this zone, and cord operated switches, only SELV switches are permitted.
- Provided they are suitable, fixed items of current using equipment such as:
  Showers
  Shower pumps
  Towel rails
  Luminaires, etc.
- Equipment designed for use in this zone must be to at least IPX4, or IPX5, where water jets are likely to be used for cleaning purposes.

Zone 2

This extends 600 mm beyond zone 1 and to a height of 2.25 m above floor level (Figure 7.3).

Note

- Other than switches and controls of equipment specifically designed for use in this zone, and cord operated switches, only SELV switches are permitted.
- Equipment designed for use in this zone must be to at least IPX4, or IPX5 where water jets are likely to be used for cleaning purposes.
- For showers without basins there is no zone 2, just an extended zone 1.
- Socket outlets other than SELV may not be installed within 3 m of the boundary of zone 1.

Additional protection

All low voltage circuits in or passing through these locations must be protected by a 30 mA RCD, or less, that operates within 40 ms at $5 \times I_{\triangle n}$.

Supplementary equipotential bonding

Supplementary bonding may be established connecting together the cpcs, exposed and extraneous conductive parts within the location.

FIGURE 7.3 Zones 0, 1 and 2.

Such extraneous conductive parts will include:

- metallic gas, water, waste and central heating pipes
- metallic structural parts that are accessible to touch
- metal baths and shower basins.

This bonding may be carried out inside or outside the location, preferably close to the entry of the extraneous conductive parts to the location.

However, this bonding may be omitted if the premises has a protective earthing and automatic disconnection system in place; all extraneous conductive parts of the locations are connected to the protective bonding and all circuits are residual current device (RCD) protected (which they have to be anyway!!).

Electric floor units may be installed below any zone provided that they are covered with an earthed metal grid or metallic sheath and connected to the protective conductor of the supply circuit.

BS 7671 SECTION 702: SWIMMING POOLS

In a similar fashion to bathrooms and shower rooms, etc., swimming pool locations are also divided into zones 0, 1 and 2:

Zone 0 is in the pool/basin or fountain.

Zone 1 extends 2.0 m horizontally from the rim of zone 0 and 2.5 m vertically above it regardless of the pool being above or below ground level. If there are diving boards, shutes or viewing galleries, etc. the height extends to a point 2.5 m from their top surface and 1.5 m horizontally either side of such shutes, etc.

Zone 2 extends a further 1.5 m horizontally from the edge of zone 1 and 2.5 m above ground level.

Now, what can we install in these zones?

Zones 0 and 1

Protection against shock

Only SELV to be used.

Wiring systems

Only systems supplying equipment in these zones are permitted. Metal cable sheaths or metallic covering of wiring systems shall be connected to the supplementary equipotential bonding. Cables should preferably be enclosed in PVC conduit.

Switchgear, control gear and socket outlets

None permitted except for locations where there is no zone 2. In this case, a switch or socket outlet with an insulated cap or cover may be installed beyond 1.25 m from the edge of zone 0 at a height of not less than 300 mm. Additionally, the circuits must be protected by:

1. SELV or
2. Automatic disconnection using a 30 mA RCD or
3. Electrical separation.

Equipment

Only that which is designed for these locations.

Other equipment may be used when the pool/basin is not in use (cleaning, maintenance, etc.) provided the circuits are protected by:

1. SELV or
2. Automatic disconnection using a 30 mA RCD or
3. Electrical separation.

Socket outlets and control devices should have a warning notice indicating to the user that they should not be used unless the location is unoccupied by persons.

Zone 2 (there is no zone 2 for fountains)

Switchgear and control gear

Socket outlets and switches, provided they are protected by:

1. SELV or
2. Automatic disconnection using a 30 mA RCD or
3. Electrical separation.

IP rating of enclosures

Zone 0 IPX8 (submersion)

Zone 1 IPX4 (splashproof) or IPX5 (where water jets are used for cleaning)

Zone 2 IPX2 (drip proof) indoor pools

IPX4 (splashproof) outdoor pools

IPX5 (where water jets are used for cleaning).

Supplementary bonding

All extraneous conductive parts in zones 0, 1 and 2 must be connected by supplementary bonding conductors to the protective conductors of exposed conductive parts in these zones.

BS 7671 SECTION 703: HOT AIR SAUNAS

Once again a zonal system, that is, 1, 2 and 3, has been used as per Figure 7.4. In this case the zones are based on temperature.

Sauna room

FIGURE 7.4 Sauna room zones.

Additional protection

All circuits in the location should have additional protection against shock by 30 mA RCDs except sauna heater circuits unless recommended by the manufacturer.

Wiring systems

It is preferred that the wiring systems for the sauna will be installed outside. However, any wiring inside should be heat resisting and any metal sheaths or conduit must be inaccessible in normal use.

Equipment

All should be at least IPX4 and IPX5 if water jets are to be used for cleaning:

Zone 1 only the sauna equipment
Zone 2 no restriction regarding temperature resistance
Zone 3 must be suitable for 125°C and cable sheaths for 175°C.

Switchgear, control gear and accessories

Only that which is associated with the sauna heater equipment may be installed in zone 2 and in accordance with the manufacturer's instructions. All others should be outside.

BS 7671 SECTION 704: CONSTRUCTION SITES

Not as complicated as one may think. The only areas that require special consideration are where construction work is being carried out, not site huts, etc.

So, let us keep all this as simple as possible. Clearly, construction sites are hazardous areas and in consequence the shock risk is greater.

Protection

For socket outlet circuits of rating up to and including 32 A and circuits supplying equipment of rating up to and including 32 A, the means of protection shall be:

1. Reduced low voltage (preferred for portable hand tools and hand lamps up to 2 kW)
2. Automatic disconnection of supply with additional protection by 30 mA RCDs
3. Electrical separation
4. SELV or PELV (SELV being preferred for portable hand lamps in damp locations).

For socket outlet circuits rated above 32 A, a 500 mA RCD is required.

External influences

These are not addressed in BS 7671, presumably as there are so many different possibilities. So common sense must prevail and equipment used with an appropriate degree of protection in accordance with the severity of the influence.

Wiring systems

Apart from some requirements for flexible cables, the only comment relates to ensuring that cables that pass under site roads, etc. are protected against mechanical damage.

Isolation and switching

An Assembly for Construction Sites (ACS), which is basically the main intake supply board, should comprise a lockable isolator and, for current using equipment and socket outlets:

1. Overcurrent devices
2. Fault protective devices
3. Socket outlets if required.

Plugs and sockets/cable couplers

All should be to BS EN 60309-2.

BS 7671 SECTION 705: AGRICULTURAL AND HORTICULTURAL LOCATIONS

The requirements apply only to locations that do not include the main farmhouse outside of which the environment is hazardous and where, of course, livestock is present (animals are susceptible to lethal shock levels at 25 V a.c.).

Protection

Protection against shock may be provided by:

1. Automatic disconnection of supply with additional RCD protection for:
   (a) Final circuits supplying socket outlets rated at 32 A or less (30 mA)
   (b) Final circuits supplying socket outlets rated more than 32 A (100 mA)
   (c) All other circuits (300 mA).
2. SELV or PELV.

Protection against thermal effects:

1. Heating appliances should be mounted at appropriate distances from combustible materials and livestock, with radiant heaters at a minimum distance of 0.5 m.
2. For fire protection an RCD rated at 300 mA or less should be used.

Additional protection

All low voltage circuits of these locations must be protected by a 30 mA RCD or less, that operates within 40 ms at $5 \times I_{\triangle n}$.

Supplementary bonding

Wherever livestock is housed, supplementary bonding must be carried out connecting all exposed and extraneous conductive parts that can be touched by livestock. All metal grids in floors must be connected to the supplementary equipotential bonding.

External influences

1. All equipments must be to at least IP44 and luminaires exposed to dust and moisture ingress, IP54.
2. Appropriate protection for socket outlets where influences are greater than AD4, AE3 and/or AG1.
3. Appropriate protection where corrosive substances are present.

Diagrams

The user of the installation should be provided with plans and diagrams showing the location of all equipment, concealed cable routes, distribution and the equipotential bonding system.

Wiring systems

Any, just as long as it is suitable for the environment and fulfils the required minimum degrees of protection.

A high impact PVC conduit/trunking system would be appropriate in many cases as it is not affected by corrosion, is rodent proof and has no exposed conductive parts. However, the system would be designed to suit the particular environmental conditions.

Wiring systems should be erected so as to be, where possible, inaccessible to livestock. Overhead lines should be insulated and, where vehicles/mobile equipment are used, underground cables should be at least 0.6 m deep and mechanically protected and 1.0 m deep in arable land.

Self-supporting suspended cables should be at a height of at least 6 m.

Switchgear and control gear

Whatever! As long as it is suitable for the conditions and that emergency switching is placed in a position inaccessible to livestock and can be accessed in the event of livestock panic (Stampede!!!).

BS 7671 SECTION 706: CONDUCTIVE LOCATIONS WITH RESTRICTIVE MOVEMENT

These are very rare locations which could include metal tanks, boilers, ventilation ducts, etc., where access is required for maintenance, repair or inspection. Bodily movement will be severely restricted and in consequence such areas are extremely dangerous.

This section deals with the installation inside the location and the requirements for bringing in accessories/equipment from outside.

For fixed equipment in the location, one of the following methods of protection shall be used:

1. Automatic disconnection of supply but with additional supplementary bonding
2. The use of Class II equipment backed up by a 30 mA RCD
3. Electrical separation
4. SELV.

For hand-held lamps and tools and mobile equipment, SELV or electrical separation should be used.

BS 7671 SECTION 708: CARAVAN AND CAMPING PARKS

We drive into a caravan/camping park for our holiday and need to connect to a supply of electricity for all our usual needs. This is accommodated by the provision of suitably placed socket outlets, supplied via distribution circuits.

External influences

Equipment should have at least the following protection codes:

1. IPX4 for the presence of splashes (AD4)
2. IP3X for the presence of small objects (AE2)
3. IK08 for the presence of high severity mechanical stress (AG3)
   (the IK codes are for impact and 08 is an impact of 5 joules).

Wiring systems

The distribution circuits are erected either underground or overhead:

1. Underground cable (preferred) should be suitably protected against mechanical damage, tent pegs, steel spikes, etc. and at a depth of not less than 0.6 m.
2. If overhead, then 6 m above ground where there is vehicle movement and 3.5 m elsewhere.

Switchgear and socket outlets

1. Supply equipment should be adjacent to, or within 20 m of, the pitch.
2. Socket outlets should be: to BS EN 60309-2; IP44, at between 0.5 m and 1.5 m above ground, rated not less than 16 A and have individual overcurrent and 30 mA RCD protection.
3. If the supply has a PME terminal, the protective conductor of each socket needs to be connected to an earth rod.

BS 7671 SECTION 709: MARINAS

This location is basically a camping park for boats and has similar requirements to those of caravan/camping parks.

It is where you arrive in your 40 ft 8 berth cruiser (some hope) looking for a place to park!

However, the environment is a little more harsh than the caravan park due to the possibilities of corrosion, mechanical damage, structural movement and flammable fuels, together with the increased risk of electric shock.

External influences

Due to the harsh conditions mentioned the classification of influences would include:

AD water
AE solid foreign bodies
AF corrosion and
AG impact.

Wiring systems

Distribution circuits, like those in caravan parks, can be either underground or overhead, as well as PVC covered mineral insulated, cables in cable management systems, etc.

However, overhead cables on or incorporating a support wire, cables with aluminium conductors or mineral insulated cables shall not be installed above a jetty or pontoon, etc.

Underground cables should have additional mechanical protection and be installed 0.5 m deep.

Overhead cables should be at the same heights as in caravan parks.

Isolation, switching and socket outlets

Generally the same as caravan parks.

Socket outlets should be installed not less than 1 m above the highest water level except that for floating pontoons, walkways, etc. this height may be reduced to 300 mm.

BS 7671 SECTION 710: MEDICAL LOCATIONS

These locations include hospitals, private clinics, dental practices, health-care centres etc. Installations in such establishments are inevitably complex and hence only a brief overview is considered here.

Rooms in medical locations are divided into groups 0, 1, and 2. Group 0 are massage rooms, group 1 are generally treatment rooms, physiotherapy, hydrotherapy radiology, etc. Group 2 are operating and intensive care, etc. rooms.

Main points:

- Safety sources feeding essential services must be provided i.e. standby systems
- Where required, in groups 1 and 2 only type A or B RCDs are permitted.
- For TN systems, final circuits up to 63A shall have protection by 30 mA or less RCDs.
- IT systems shall be used for equipment and systems in group 2 locations intended for life support and surgical applications, etc.
- Supplementary equipotential bonding shall be provided in each medical location of group 1 and 2.
- Unwanted tripping of 30 mA or less RCDs must be taken into consideration.
- For IT systems in group 2, socket outlets supplying medical equipment must be un-switched and fitted with a supply indicator.
- For initial verification, the additional functional tests of 'insulation monitoring devices' and 'overload monitoring' together with the verification of the integrity of the supplementary bonding must be carried out.

BS 7671 SECTION 711: EXHIBITIONS, SHOWS AND STANDS

This section deals with the protection of the users of temporary structures erected in or out of doors and is typical of antique fairs, motorbike shows, arts and craft exhibitions, etc.

It does not cover public or private events that form part of entertainment activities, which are the subject of BS 7909.

External influences

None particularly specified. Clearly they must be considered and addressed accordingly.

Wiring

Armoured or mechanically protected cables where there is a risk of mechanical damage.

Cables shall have a minimum conductor size of $1.5\,mm^2$.

Protection

Against shock:

1. Supply cables to a stand or unit, etc. must be protected at the cable origin by a time-delayed RCD of residual current rating not exceeding 300 mA.
2. All socket outlet circuits not exceeding 32 A and all other final circuits, excepting emergency lighting, shall have additional protection by 30 mA RCDs.
3. Any metallic structural parts accessible from within the unit stand, etc. shall be connected by a main protective bonding conductor to the main earthing terminal of the unit.

Against thermal effects:

1. Clearly in this case all luminaires, spot lights, etc. should be placed in such positions as not to cause a build-up of excessive heat that could result in fire or burns.

Isolation

Every unit, etc. should have a readily accessible and identifiable means of isolation.

Inspection and testing

Tongue in cheek here! Every installation **should** be inspected and tested on site in accordance with Part 6 of BS 7671.

BS 7671 SECTION 712: SOLAR PHOTOVOLTAIC (PV) SUPPLY SYSTEMS

These are basically solar panels generating d.c. which is then converted to a.c. via an invertor. Those dealt with in BS 7671 relate to those systems that are used to 'top up' the normal supply.

There is a need for consideration of the external influences that may affect cabling from the solar units outside to control gear inside.

There must be protection against overcurrent and a provision made for isolation on both the d.c. and a.c. sides of the invertor.

As the systems can be used in parallel with or as a switched alternative to the public supply, reference should be made to Chapter 55 of BS 7671.

BS 7671 SECTION 714: OUTDOOR LIGHTING INSTALLATIONS

This section applies, for example, to lighting installations for, roads, car parks, monuments, road signs, town plans etc. but, interestingly not to a luminaire fixed to the outside wall of a premises and fed from the internal wiring.

The method of protection against electric shock for these installations is the standard Automatic Disconnection of Supply with a maximum disconnection time of 5 s.

For items such as town plans, telephone kiosks, bus shelters etc., 30 mA or less RCD protection is required.

External influences should be addressed and equipment should be at least IP33 rated.

BS 7671 SECTION 715: EXTRA LOW VOLTAGE LIGHTING INSTALLATIONS

In this case, basic and fault protection is provided by SELV or PELV fed from an isolating transformer and where bare conductors are used such as in track systems the maximum voltage is 25 V a.c. or 60 V d.c.

Protection against fire risk from transformers and from short circuits must be provided.

A maximum voltage drop of 5% between the transformer and the furthest luminaire is acceptable.

BS 7671 SECTION 717: MOBILE OR TRANSPORTABLE UNITS

Medical facilities units, mobile workshops, canteens, etc. are the subject of this section. They are self-contained with their own installation and designed to be connected to a supply by, for instance, a plug and socket.

The standard installation protective measures against shock are required with the added condition that the automatic disconnection of the supply should be by means of an RCD.

Also, all socket outlets for the use of equipment outside the unit should have additional protection by 30 mA RCDs.

The supply cable should be HORN-F, oil and flame resistant heavy duty rubber with a minimum copper conductor size of 2.5 mm^2.

Socket outlets outside should be to a minimum of IP44.

BS 7671 SECTION 721: CARAVANS AND MOTOR CARAVANS

These are the little homes that people tow behind their cars or that are motorized, not those that tend to be located on a fixed site. It would be unusual for the general Electrical Contractor to wire new, or even rewire old units. How many of us ever rewire our cars? In consequence, only the very basic requirements are considered here.

Protection

These units are small houses on wheels and subject to the basic require-ments of protection against shock and overcurrent. Where automatic dis-connection of supply is used this must be provided by a 30 mA RCD.

Wiring systems

The wiring systems should take into account the fact that the structure of the unit is subject to flexible/mechanical stresses and, therefore, our common flat twin and three-core cables should not be used.

Inlets

Unless the caravan demand exceeds 16 A, the inlet should conform to the following:

(a) To BS EN 60309-1 or 2 if interchangeability is required
(b) Not more than 1.8 m above ground level
(c) Readily accessible and in a suitable enclosure outside the caravan
(d) Identified by a notice that details the nominal voltage, frequency and rated current of the unit.

Also, inside the caravan, there should be an isolating switch and a notice detailing the instructions for the connection and disconnection of the elec-tricity supply and the period of time between inspection and testing (3 years).

General

Accessories and luminaires should be arranged such that no damage can occur due to movement, etc.

There should be no compatibility between sockets of low and extra low voltage.

Any accessory exposed to moisture should be IP55 rated (jet proof and dust proof).

BS 7671 SECTION 722: ELECTRIC VEHICLE CHARGING INSTALLATIONS

There are four methods of electric vehicle (EV) charging known as Modes and are as follows:

Mode 1

Connecting the EV to a standard earthed socket outlet not exceeding 16 A e.g. BS 1363 or BS EN 60309.

Mode 2

As Mode 1 but up to 32 A and a special cable that has a connector/control box with an extra pin at the EV end. This is known as a 'pilot control function' and ensures that current can only flow if the earthing is intact. There must also be RCD protection between the EV and the control box.

Mode 3

This requires the use of dedicated EV supply equipment (EVSE) which incorporates the pilot control function etc. The connectors/plugs are to have control and signal pins which ensure the EVSE socket cannot be energised if an EV is not connected.

Mode 4

In this instance, the a.c. mains is converted to d.c. via a charger incorporating the pilot control function. The EV is connected to this charging unit.

EV charging point equipment outdoors should be at least IP44 rated.

BS 7671 SECTION 729: OPERATING AND MAINTENANCE GANGWAYS

Such gangways are likely to be found in restricted areas that are typical of switchrooms, etc., where protection against contact with live electrical parts of equipment is provided by barriers or enclosures or obstacles, the latter having to be under the control of a skilled persons.

Main points:

- Restricted areas must be clearly and visibly marked.
- Access to unauthorized persons is not permitted.

- For closed restricted areas, doors must allow for easy evacuation without the use of a key or tool.
- Gangways must be wide enough for easy access for working and for emergency evacuation.
- Gangways must permit at least a 90° opening of equipment doors etc.
- Gangways longer than 10 m must be accessible from both ends.

BS 7671 SECTION 740: AMUSEMENT DEVICES, FAIRGROUNDS, CIRCUSES, ETC.

This is not an area that is familiar to most installation electricians and hence will only be dealt with very briefly.

The requirements of this section are very similar to those of Section 711 Exhibitions, shows, etc. and parts of Section 706 Agricultural locations (because of animals) regarding supplementary bonding.

For example, additional protection by 30 mA is required for:

1. Lighting circuits, except those that are placed out of arm's reach and not supplied via socket outlets.
2. All socket outlet circuits rated up to 32 A.
3. Mobile equipment supplied by a flexible cable rated up to 32 A.

Automatic disconnection of supply must be by an RCD.

Equipment should be rated to at least IP44.

The installation between the origin and any equipment should be inspected and tested after each assembly on site.

BS 7671 SECTION 753: FLOOR AND CEILING HEATING SYSTEMS

Systems referred to in this section are those used for thermal storage heating or direct heating.

Protection

Against shock:

1. Automatic disconnection of supply with disconnection achieved by 30 mA RCD.
2. Additional protection for Class II equipment by 30 mA RCDs.
3. Heating systems provided without exposed conductive parts shall have a metallic grid of spacing not more than 300 mm installed on site above a floor system or below a ceiling system and connected to the protective conductor of the system.

Against thermal effects:

1. Where skin or footwear may come into contact with floors the temperature shall be limited, for example to 30°C.
2. To protect against overheating of these systems the temperature of any zone should be limited to a maximum of 80°C.

External influences

Minimum of IPX1 for ceilings and IPX7 for floors.

The designer must provide a comprehensive and detailed plan of the installation which should be fixed on or adjacent to the system distribution board.

Questions

1. When is the space under a bath or shower outside of all the zones?
2. What is the IP rating for equipment in zone 1 of a swimming pool where jets are use for cleaning purposes?
3. What is the maximum height above floor level of zone 2 of a sauna?
4. What rating of RCD is required for a 63 A socket outlet on a construction site?
5. How deep should a cable be buried in arable ground on a farm?
6. What is the maximum rating of an RCD used in conjunction with Class II fixed equipment in a conductive location with restricted movement?
7. What is the impact code for equipment subject to (AG 3) mechanical stress on a caravan park?
8. What is the minimum height of a socket outlet above the highest water level for a walkway in a marina?

9. What is the room Group reference for an operating theatre in a hospital?
10. What rating and type of RCD should be provided at the origin of the supply cable to a stand in an exhibition?
11. For a PV installation, where should inverter isolation be provided?
12. Which socket outlets associated with mobile units should be 30 mA, or less, protected?
13. What is the maximum height above ground level for the inlet to a caravan?
14. Above what distance should a gangway be accessible from both ends?
15. What is the minimum IP rating for equipment in a fairground?
16. What is the minimum IP rating for ceiling heating systems?
17. What is the maximum disconnection time for outdoor lighting installations?
18. What is the maximum a.c. voltage for an extra low voltage lighting system with bare conductors?

Answers

1. When accessible only by a tool
2. IPX5
3. 1 metre
4. 500 mA
5. 1 metre
6. 30 mA or less
7. IK08
8. 300 mm
9. Group 2
10. 300 mA time delayed or 'S' type
11. On both the a.c. and the d.c. side
12. Those used for equipment outside the unit
13. 1.8 m
14. 10 m
15. IP44
16. IPX1
17. 5 s
18. 25 V

17th Edition: Examination Question Allocation and Sample Paper

TYPICAL CITY & GUILDS STYLE MULTIPLE CHOICE QUESTION PAPER

The examination comprises 60 questions to be answered in two hours. All papers are divided into eight sections representing the seven Parts and the appendices of BS 7671.

Each section has a set number of questions as shown in the table below, and questions follow through the Regulations from Part 1 to the appendices.

Part	Title	Number of Questions	Question Numbers
1	Scope and Object etc	4	*1 to 4*
2	Definitions	3	*5 to 7*
3	Assessment of Gen. Char.	5	*8 to 12*
4	Protection for Safety	14	*13 to 26*
5	Selection of Equipment	15	*27 to 41*
6	Testing	6	*42 to 47*
7	Special Locations	10	*48 to 57*
8	Appendices	3	*58 to 60*

SAMPLE PAPER

1. BS 7671:2008 applies to electrical installations in
   a. public premises
   b. ships
   c. aircraft
   d. mines and quarries

2. Protection against direct contact may be provided by
   a. fault protection
   b. overcurrent protection
   c. basic protection
   d. protection against thermal effects

3. The selection of protective equipment would NOT include consideration of the effects of
   a. overcurrent
   b. equipment costs
   c. earth fault current
   d. undervoltage

4. The interval between initial verification and the first periodic inspection of an installation is determined by
   a. the installer
   b. the client
   c. the Local Authority Building Control
   d. the designer

5. A cable support consisting of a series of transverse supporting elements rigidly fixed to main longitudinal supporting members is
   a. cable ladder
   b. cable ducting
   c. cable tray
   d. cable trunking

6. The magnitude of current to be carried by a circuit in normal service is
   a. overload current
   b. short circuit current
   c. protective conductor current
   d. design current

7. Protection against electric shock under single fault conditions is
   a. emergency stopping
   b. fault protection
   c. basic protection
   d. protective separation

8. When assessing the maximum demand of an installation account may be taken of
   a. discrimination
   b. diversity
   c. arrangement of live conductors

   d. type of earthing arrangement

9. Every installation shall be divided into circuits in order to
   a. prevent insulation faults
   b. reduce electricity costs
   c. to meet the requirements of the NICEIC
   d. mitigate the effects of EMI

10. Electrical equipment in, for example, kiosks or shops within shops need to be assessed with regards to
    a. maintainability
    b. continuity of service
    c. compatibility
    d. safety services

11. Which of the following is NOT recognised as a source for a safety service.
    a. Storage batteries
    b. The normal DNO supply
    c. Primary cells
    d. Independent generator sets

12. An assessment for continuity of service must include
    a. details of the installer
    b. details of the supplier
    c. number of circuits
    d. selection of cable type

13. Automatic disconnection of supply provides protection against
    a. overload
    b. electric shock
    c. undervoltage
    d. thermal effects

14. Basic insulation of live parts or barriers or enclosures provide
    a. impact protection
    b. thermal protection
    c. basic protection
    d. fault protection

15. A 20 A final circuit fed by a 230 V a.c. TT system must disconnect, in the event of an earth fault, within a maximum time of
    a. 0.1 s
    b. 0.2 s
    c. 0.4 s
    d. 1 s

16. A 13 A socket outlet does NOT need to be additionally protected by an RCD provided it is
    a. not for use outdoors
    b. only to be used by ordinary persons
    c. splashproof
    d. specifically labelled for a particular item

17. The maximum earth loop impedance for a distribution circuit protected by a 32A BS 88-3 fuse is
    a. 1.6 Ω
    b. 1.7 Ω
    c. 1.37 Ω
    d. 0.91 Ω

18. The maximum rating of an RCD required to protect a 230 V TT installation where the earth fault loop impedance is 150Ω is
    a. 30 mA
    b. 100 mA
    c. 300 mA
    d. 500 mA

19. The maximum disconnection time for circuits supplied by a reduced low voltage system is
    a. 0.2 s
    b. 0.4 s
    c. 1 s
    d. 5 s

20. A 16A BS EN 60898 Type B circuit breaker used to protect a three-phase reduced low voltage circuit is subject to a maximum earth fault loop impedance value of
    a. 0.33 Ω
    b. 0.38 Ω
    c. 0.65 Ω
    d. 0.75 Ω

21. For the protective measure of electrical separation, the voltage of a separated circuit must not exceed
    a. 230 V
    b. 400 V
    c. 500 V
    d. 1000 V

22. Which of the following sources of supply would be suitable for SELV and PELV circuits?
    a. Public supply at 230 V a.c.
    b. Private generator at 110 V a.c.
    c. A safety isolating transformer
    d. A low voltage d.c. supply

23. Which of the following formulae is used where there is doubt regarding the effectiveness of supplementary bonding?
    a. $R \geq 50 \text{ V}/I_a$
    b. $R \leq 50 \text{ V}/I_n$
    c. $R \leq 50 \text{ V}/I_a$
    d. $R \leq 50 \text{ V}/Iz$

24. In order to protect against burns, a non-metallic electrical part intended to be

touched but not hand held
should not attain a surface
temperature in excess of

a. 65 °C
b. 70 °C
c. 80 °C
d. 90 °C

25. The disconnection time for a
16.0 mm^2 conductor having a
'k' factor of 115 and carrying a
fault current of 13kA is

a. 0.0125 s
b. 0.02 s
c. 0.14 s
d. 2.26 s

26. The minimum impulse
withstand voltage for a 230 V
electricity meter is

a. 1.5 kV
b. 2.5 kV
c. 4 kV
d. 6 kV

27. A conductor marked green-
and-yellow throughout
its length with additional
blue markings at the
termination is a

a. line conductor
b. neutral conductor
c. PEN conductor
d. circuit protective
   conductor

28. The notice relating to an
RCD requires the device to be
operated via its test button

a. monthly
b. quarterly
c. half yearly
d. annually

29. One method of protecting a
cable, that passes through a
ceiling joist from damage, is to
install it at a vertical distance
from the top or bottom of the
joist of at least

a. 10 mm
b. 25 mm
c. 30 mm
d. 50 mm

30. The derating factor for a
cable surrounded by thermal
insulation for 200 mm is

a. 0.5
b. 0.51
c. 0.63
d. 0.78

31. Where underground
telecommunication and power
cables cross, the minimum
clearance to be maintained
between them is

a. 50 mm
b. 100 mm
c. 200 mm
d. 0.5 m

32. The magnetic circuit of an RCD shall enclose
    a. all live conductors
    b. all line conductors
    c. all circuit conductors
    d. PEN conductors

33. Where an RCD is used for protection against fire, it shall have a maximum rating of
    a. 30 mA
    b. 100 mA
    c. 300 mA
    d. 500 mA

34. Which of the following devices may **only** be used for functional switching?
    a. A BS EN 60898 circuit breaker
    b. A 16A plug and socket-outlet
    c. A cooker control unit switch
    d. A device with semi-conductors

35. A firefighter's switch should be mounted above ground level at a height of
    a. 2.75 m with the switch OFF position at the top
    b. 2.75 m with the switch ON position at the top
    c. 2.25 m with the switch OFF position at the top
    d. 2.75 m with the switch OFF position at the bottom

36. The minimum size of a buried copper earthing conductor not protected against mechanical damage or corrosion is
    a. 10 mm^2
    b. 16 mm^2
    c. 25 mm^2
    d. 50 mm^2

37. The minimum size of a protective conductor with a 'k' factor of 143 which disconnects in 0.3 s at a fault current of 800A is
    a. 2.5 mm^2
    b. 4.0 mm^2
    c. 6.0 mm^2
    d. 10.0 mm^2

38. A distribution board containing circuits with high protective conductor currents shall have the information regarding these circuits positioned so as to be visible to
    a. the Distribution Network Operator
    b. the user of the installation
    c. a person modifying or extending a circuit
    d. an inspector

39. The earthing conductor for an installation supplied from a TN-S system is 16 mm². The minimum size of a main protective bonding conductor is
a. 6 mm²
b. 10 mm²
c. 16 mm²
d. 25 mm²

40. An installation and its generator are not permanently fixed. In a TN, TT or an IT system an RCD shall be installed with a maximum rating of
a. 500 mA
b. 300 mA
c. 100 mA
d. 30 mA

41. Control equipment incorporating protection against overload shall be provided for every motor having a rating exceeding
a. 100 W
b. 0.37 kW
c. 37 W
d. 10 kW

42. The inspection of an installation is NOT made to ensure
a. compliance with section 511 of BS 7671

b. that equipment is correctly selected and erected
c. that the requirements of the ESQCR are met
d. that it is not visible damaged or defective

43. The minimum value of insulation resistance and the applied test voltage for a 400 V circuit is
a. 1.0MΩ at 500 V a.c.
b. 1.0MΩ at 500 V d.c.
c. 0.5MΩ at 250 V d.c.
d. 1.0MΩ at 1000 V d.c.

44. Where surge protective devices cannot be disconnected for insulation resistance testing, the minimum value of insulation resistance and the applied test voltage are
a. 1.0 MΩ at 500 V a.c.
b. 1.0 MΩ at 500 V d.c.
c. 0.5 MΩ at 250 V d.c.
d. 1.0 MΩ at 250 V d.c.

45. The correct connection of socket-outlets and similar accessories is established by
a. functional testing
b. polarity testing
c. insulation resistance testing
d. RCD testing

46. An installation is to have the following tests carried out:
1. polarity. 2. cpc continuity 3. earth electrode resistance. 4. insulation resistance.
The correct sequence for these tests is
    a. 1, 2, 3, 4
    b. 4, 2, 3, 1
    c. 3, 1, 4, 2
    d. 2, 4, 1, 3

47. The verification of voltage drop may be achieved by
    a. use of a voltmeter
    b. measuring the circuit impedance
    c. measuring the prospective fault current
    d. functional testing

48. A 13 A socket-outlet may be installed in a bathroom if it is installed beyond the boundary of zone 1 by a horizontal distance of
    a. 0.60 mm
    b. 0.50 mm
    c. 2.25 m
    d. 3.0 m

49. The external influence code for zone 0 of a swimming pool is
    a. AG3
    b. AE6
    c. AD8
    d. AF2

50. Cables buried in arable land on a farm should be at a minimum depth of
    a. 0.5 m
    b. 0.6 m
    c. 1.0 m
    d. 6 m

51. Electrical equipment installed outside in a caravan park shall have a mechanical impact code of at least
    a. IK01
    b. IK03
    c. IK06
    d. IK08

52. Socket-outlets in a marina shall be protected
    a. in groups by RCDs
    b. individually by RCDs and overcurrent devices
    c. residual current monitoring devices
    d. BS 3036 fuses only

53. Temporary installations of exhibitions should inspected and tested
    a. after each assembly on site
    b. before disassembly

c. at intervals during use

d. only after a fault has occurred

54. The flexible cable for connecting a mobile workshop to the supply shall have a minimum cross sectional area of

a. 1.5 mm² copper

b. 1.5 mm² aluminium

c. 2.5 mm² copper

d. 4.0 mm² copper

55. A periodic inspection for a frequently used caravan should be carried out

a. monthly

b. quarterly

c. annually

d. every three years

56. All final circuits in a amusement park for lighting and socket-outlets and mobile equipment up to 32 A must have additional protection by

a. BS 88 fuses

b. 30 mA or less RCDs

c. 300 mA or less RCDs

d. 32 A BS EN 60898 circuit breakers

57. Heating unit in ceilings shall have a degree of protection of at least

a. IP4X

b. IP3X

c. IP2X

d. IP1X

58. A 16 A BS 88-3 fuse carrying a fault current of 100 A will disconnect in

a. 1 s

b. 2 s

c. 3 s

d. 4 s

59. The maximum permissible voltage drop for a 400 V motor circuit is

a. 20 V

b. 11.5 V

c. 12 V

d. 6.9 V

60. The maximum value of earth fault loop impedance for a 32 A BS EN 60898 type B circuit breaker, for comparison with test values is

a. 1.37 Ω

b. 1.09 Ω

c. 0.68 Ω

d. 0.54 Ω

Answers to Multiple Choice Test

1 a	11 b	21 c	31 b	41 b	51 d
2 c	12 c	22 c	32 a	42 c	52 b
3 b	13 b	23 c	33 c	43 a	53 a
4 d	14 c	24 c	34 d	44 d	54 c
5 a	15 b	25 b	35 a	45 b	55 c
6 d	16 d	26 d	36 c	46 d	56 b
7 b	17 a	27 c	37 b	47 b	57 d
8 b	18 c	28 b	38 c	48 d	58 c
9 d	19 d	29 d	39 b	49 c	59 a
10 c	20 d	30 c	40 d	50 c	60 b

Index

Note: page numbers in italics refer to figures; page number in bold refer to tables

A

adiabatic equation 75, 77–8
agricultural locations 118–20
aluminium conductors: resistance for
   various cross-sections **31**
ambient temperature 65, 68
amusement devices 129
arm's reach 39, 48
automatic disconnection 22, 48
auxiliary circuits 63

B

barriers 39, 47–8
basic protection 13, 39, 47–8
bathrooms 110–13
bonding conductors 13, 22, 36–7, 91,
   92
breaking capacities 52, **53–4**
Building Regulations 2010 Part 'P' 8–10

C

cables: construction 40–1; current-
   carrying capacity 71–2; grouping 68,
   69–70; installation 78–9, *80*, *81*, *82*,
   83; rating factors (CF) 68–71;
   sheathing 23, 42; size selection 72–3,
   75, 85–7
camping parks 120–1
caravan parks 120–1
caravans 126–7
ceiling heating 129–30
certification 8–9
circuit breakers: breaking capacities 52,
   53; operation 54–8; *see also* residual
   current devices (RCDs)
circuit design 65–87; cable ratings 68–71;
   cable selection 72–3; design current
   67; design procedure 66; example
   calculations 79–87; nominal setting
   of protection 67; shock risk 74;
   thermal constraints 74–9; voltage drop
   calculation 73–4

circuit protective conductors (cpcs) 13,
   22, 39, 65; continuity testing 92–3; size
   selection 74–7
circuses 129
City & Guilds style questions 133–41
Class II equipment 39, 48
competent persons 8–9
conductor cross-sections and resistance
   28, **31**, **91**
confined spaces 120
construction sites 116–18
consumer units 8, 35–6, 43
continuity testing 90–6
cooker circuits 35, **81**, 91
copper conductors: resistance for various
   cross-sections **28**, **31**
corrosion protection 42
current-carrying capacity 65, 71–2

D

design current 39, 65, 67
design procedure 66
discrimination **57**, 58–9
double insulation 48

E

earth 13, 15–18
earth conductors *see* circuit protective
   conductors (cpcs)
earth connection 18–19
earth electrodes 18–19, 89; resistance 13,
   19–21, 100
earth fault current 13, 25, 52
earth fault loop impedance 13, 25–32,
   74, 89; corrected values **102–4**; testing
   100–1
earth-free local equipotential bonding 49
earthing 13–37
earthing conductor, definition 13, 65
earthing systems 23–5
earth path 16, 18
efficiency (eff) 67

Electrical Installation Certificates (EICs) 8
Electrical Installation Condition Reports
  (EICRs) 8
electrical protection 47–60
electrical separation 49, *50*
Electricity at Work Regulations 1989 5–7
Electricity Safety, Quality and Continuity
  Regulations 2002 (ESQCR) 9
electric shock 17, 74
electric vehicle (EV) charging 128
emergency switching 61, 62
employee's duties 5, 6
employer's duties 3–4, 5
enclosures 39, 47–8
equipotential bonding 13, 17
exemptions 49–50
exhibitions 123–5
exposed conductive parts 13, 22, 39
external influences 39, **44**, 44
external loop impedance 25, 100, 105
extra low-voltage (ELV) 9, 50–1, 126
extraneous conductive parts 13, 22, 39

F

fairgrounds 129
Faraday cage 37, 49
fault current, definition 40, 65
fault protection 14, 40, 48–50
firefighters' switches 62
floor heating systems 129–30
freezers 35
functional earth 14
functional switching 62
functional testing 106
fuses: breaking capacities 52, **53–4**;
  operation 54–9
fusing factor 54–5, 68

G

gangways 128–9

H

Health and Safety at Work Act 1974 3–5
heating appliances 35, **81**
heating systems 129–30
horticultural locations 118–20
hot air saunas 115–16

I

IET Wiring Regulations (BS 7671) 1–3
IK codes 44, **46**

inspection *see* testing and inspection
installation 78–9, *80*, *81*, *82*, *83*
installers 8–9
insulation 40, 41–2, 43, 48
insulation resistance testing 97–9
insulation (thermal) and cable runs 44,
  69
IP codes 44, **45–6**, 47
isolating transformers 49
isolation 40, 61
isolation switching and control 61–3

L

leakage current 14
lighting circuits: circuit breakers **81**; high-
  temperature cables 43; polarity testing
  99; voltage drop **73**
line conductors 14
live part, definition 14, 40
local equipotential bonding 49
low-voltage (LV) 9
luminaires 43

M

main protective bonding conductors 22,
  37; continuity testing 91, *92*
marinas 121–2
mechanical maintenance 61, 62
mechanical protection 41–7
medical locations 123
metal fixings 50
metal furniture 37
metal window frames 36–7
mineral insulated cables (MICCs) 42
Minor Electrical Installation Works
  Certificates (MEIWCs) 8
mobile units 126
motor caravans 126–7
motor circuits 30–2, 62

N

nominal setting of protection 67
non-conducting locations 49
non-notifiable work 8
notifiable work 8
nuisance tripping 35

O

obstacles 40, 48
operating and maintenance gangways
  128–9

outdoor lighting 125
overcurrents 40, 65; protection 51–9
overhead cables 119, 121, 122
overloads 40, 52
overvoltage protection 59

P

PELV circuits 50–1, **98**, 126
PEN conductor 14
phase sequence 105–6
photovoltaic (pv) supply systems 125
PME (protective multiple earthing) 14, 23, 37
polarity testing 99–100
polystyrene, in contact with PVC 44
polyvinyl chloride 43; *see also* PVC cables
potential difference (PD) 15
power circuits **73**, **81**, 93–6, 99–100
power factor (PF) 67
prospective earth fault current (PEFC) 52
prospective fault current (PFC) 52, 55, 58, 105
prospective short-circuit current (PSCC) 52
protection: against corrosion 42; electrical 47–60; against ingress of solid objects, liquid and impact 44–7; against mechanical damage 41–7; against thermal effects 42–4
protective conductors 14, 22; continuity testing 90–3; resistance for various cross-sections **28**, **31**, **91**; *see also* circuit protective conductors (cpcs); main protective bonding conductors; supplementary bonding conductors
protective devices 52–9; *see also* circuit breakers; fuses; residual current devices (RCDs)
protective equipotential bonding 17
protective sheathing 42
PVC 43
PVC cables 42; embrittlement and cracking 43–4
PVC conduit 43–4

R

radial socket outlet circuits: polarity testing 99–100

radiators, bonding 36
rating factors (CF) 68–71
RCD/CB devices (RCBOs) 36, *79*, 105
'reasonably practicable' 6–7
reduced low-voltage systems 51
registered competent person 8–9
reinforced insulation 48
residual current devices (RCDs) 14, 32–6, 40, 48, 89; testing 105, 106; tripping times **105**; *see also* RCD/CB devices (RCBOs)
restricted spaces 120
ring final circuits **81**, 89; continuity testing 93–6
rod electrodes 18–19

S

sample examination paper 134–41
saunas 115–16
self-employed persons 6
SELV circuits 50–1, **98**, 126
separated extra low voltage (SELV) 50–1
sheathing 23, 42
shock path 17
shock risk *see* electric shock
short-circuit current 40, 52
shower circuits 29–30, **81**, 110–13
shows 123–5
skilled person 40
solar photovoltaic (PV) circuits 125
special locations 109–30
stands 123–5
supplementary bonding conductors 22, 36–7; continuity testing 91
supply voltages 9, *10*
swimming pools 113–15
switches 61

T

testing and inspection 89–106; continuity 90–6; earth electrode resistance 100; earth fault loop impedance 100–5; functional 106; insulation resistance 97–8; phase sequence 105–6; polarity 99–100; RCDs 105, 106; sequence 89–90; voltage drop verification 106
thermal constraints 74–9
thermal insulation 44, 69
thermal protection 42–3
TN-C-S systems 23, *24*, 27, 37

TN-S systems 23, *24*, 27
transportable units 126
tripping times **105**
TT systems 23, *24*, 27, 100

U

underground cables 119, 121, 122
undervoltage protection 60

V

voltage bands 9
voltage drop 73–4, 106
voltages 9, *10*

W

water heaters 35
water taps 36